Story of the
W and Z

Story of the W and Z

PETER WATKINS
Lecturer in Physics, University of Birmingham

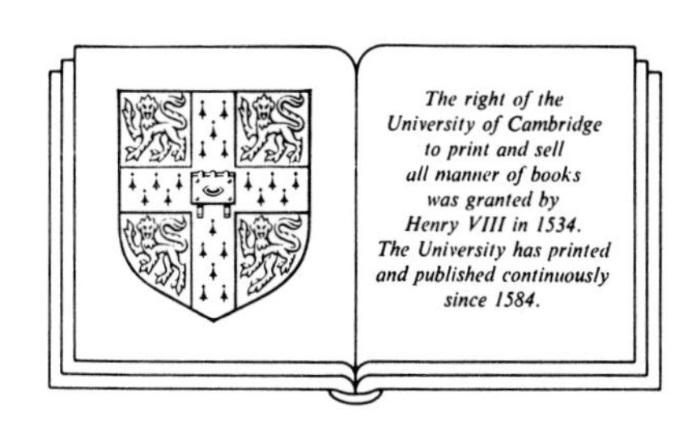

CAMBRIDGE UNIVERSITY PRESS
CAMBRIDGE
LONDON NEW YORK NEW ROCHELLE
MELBOURNE SYDNEY

Published by the Press Syndicate of the University of Cambridge
The Pitt Building, Trumpington Street, Cambridge CB2 1RP
32 East 57th Street, New York, NY 10022, USA
10 Stamford Road, Oakleigh, Melbourne 3166, Australia

First published 1986

Printed in Great Britain by the
University Press, Cambridge

British Library cataloguing in publication data

Watkins, Peter
Story of the W and Z.
1. Bosons
I. Title
539.7′21 QC793.5.B62

Library of Congress cataloguing in publication data

Watkins, P. M. (Peter Maitland), 1944–
Story of the W and Z.

Includes index.
1. W bosons. 2. Z bosons. 3. Intermediate bosons.
4. Particles (Nuclear physics)—Research—Europe.
5. European Organization for Nuclear Research.
I. Title. II. Title: W and Z.
QC793.5.B62W37 1986 539.7′21 85-17472

ISBN 0 521 26801 X hard covers
ISBN 0 521 31875 0 paperback

UP

To Anna, Ben and Clare

Contents

PART FIVE: THE FUTURE

Acknowledgements

I would like to thank all my colleagues in the UA1 collaboration and the Birmingham University Particle Physics group for many informative discussions which have helped to clarify several of the topics included in this book. The story is about the very successful antiproton–proton collider project in which the engineers and physicists at the CERN laboratory made many important contributions. The crucial roles in this project of Carlo Rubbia and Simon Van der Meer were recognised by the joint award of the 1984 Nobel Physics prize.

I am especially grateful to Martyn Corden, Glenn Cox, Goronwy Jones, David Watkins and John Wilson. They helped me enormously by reading part or, in most cases, all of the manuscript and suggesting many improvements which I have tried to incorporate. However, any remaining errors or omissions, remain my responsibility. I would also like to thank Sharon Ellis and Ann Aylott for their careful typing and Margaret Baggott for her help with the diagrams.

Finally, I thank my family for their encouragement and support during the preparation of this book.

Part One

The background

Chapter One

Introduction

This story is a first-hand account of one of the most exciting scientific searches in modern times, the recent search for the W and Z bosons. These particles play an important role in our world but most people know nothing of their existence. A prime aim of this book is to communicate to a wider audience some of the excitement of the most recent results from the study of particle physics. To appreciate these new results we have to take a journey into our imagination. Most people are aware that atoms exist and that all matter is composed of atoms, but very few know much more about this hidden world. This is very sad as we now have an enormous amount of information about the particles deep inside the atom and the forces that hold them together. Many of the results are very recent and yet the important ideas can be understood by anyone who is really interested in this subject.

Particle physics experiments are aimed at discovering the smallest particles that exist in an atom and understanding the forces that act between these particles. This study of the smallest known objects surprisingly has a number of common features with the study of our universe at large. They are both exciting frontiers of scientific research which are primarily aimed at discovering more about our natural surroundings.

This book is an account of an experiment carried out at the CERN laboratory near Geneva, which searched for the W and Z bosons in 1983. Some of the largest and most sophisticated accelerators, computers and electronic detectors in the world were used in this search. The early part of the book introduces the important properties of the atom and its nucleus and presents the experimental evidence which suggests that all matter is composed of structureless quarks and leptons. The forces of nature can, we believe, be reduced to just four fundamental forces and the properties of the gravitational, electromagnetic, weak nuclear and strong nuclear forces are discussed. This is followed by a discussion of some of the important ideas and experiments that led to the unification of the electromagnetic and weak nuclear forces. This unification predicted the existence of the W and Z bosons, each with a mass approximately 100 times greater than a proton. The following chapters describe how the idea for the antiproton–proton collider project at CERN originated and also how these W and Z bosons might be detected. In order to appreciate fully the excitement of the search, it is necessary to have some knowledge of the accelerators and detectors which are used in particle physics experiments and so these are also introduced. There were two major detectors, known as UA1 and UA2, installed at the CERN collider and these both made crucial contributions in the search for W and Z bosons. This book concentrates on the UA1 experiment as I can describe this from first-hand experience.

The story of the experiment begins with the design and construction of a 2000 tonne detector by an international collaboration. The main part of the book presents a vivid account of what it was like to work on the UA1 experiment and the tasks performed by the people involved in that collaboration. The exciting story of the preparations and running periods of the experiment are fully explained so it should be possible to get personally involved in the excitement of the search for these important new particles. The story continues with a description of the hectic analysis of results where the experimental data, recorded over many months, are carefully studied for evidence of the W and Z bosons and other new phenomena.

The book ends with a brief review of some of the most interesting current ideas and experiments in particle physics. I hope you enjoy this journey into the world of the atom and this story of a very exciting experiment. I have certainly enjoyed working on it and trying to explain some of the important and interesting ideas to you.

Chapter Two

Inside the atom

2.1 All matter is made of atoms

Our normal surroundings, although apparently solid and unchanging, conceal a vast number of rapidly moving atoms which are too small to be seen. The air that we breath is made up of a large number of atoms which are moving unseen and rapidly around us. We are also made up of atoms and so are the seemingly solid objects that we see around us. Matter exists in a variety of forms – solids, liquids and gases. These are all composed of atoms but they appear to be very different because the atoms behave rather differently in each of these states. In a solid, each atom has a permanent location and merely vibrates about this position. This stability of position means that a solid retains its shape and size. As the temperature is raised the vibrations of the atoms increase, the atoms begin to move about more freely and eventually the solid becomes a liquid. The atoms in a liquid can move; however, this remains a local motion and the liquid will flow when poured. At higher temperatures the speed of the atoms increases still further, and once the temperature reaches the boiling point then the liquid changes into a gas. In a gas the atoms are moving very rapidly and a container is required to keep them in a localised volume. The atoms move in random directions and with varying speeds, colliding with each other and the walls of the container. These collisions are responsible for the pressure exerted by the gas. In several important gases, for example oxygen, the atoms prefer to travel in pairs and these are called molecules.

All matter consists of elements, which each have slightly different properties, and all the atoms of a given element are identical. As there are over 100 different chemical elements it was a formidable task to classify them and group together those with similar properties. This was achieved by the Russian chemist Dmitri Mendeleev when he constructed his periodic table of elements. The table was organised so that similar elements occurred together and he also predicted the properties of three extra elements which seemed to be missing from the pattern. Over the next 30 years these missing elements were discovered by scientists in France, Germany and Scandinavia and are now known as gallium, germanium and scandium. This was a triumph in the classification of atoms and was the basis for dramatic progress in chemistry which deals with the interactions between atoms.

The size of atoms

We can directly observe objects as small as one-tenth of a millimetre, or 10^{-4} metres, without any extra equipment. Once we use an optical microscope we can see much smaller objects, perhaps down to 10^{-7} metres, 1000 times smaller. This means that we can study living cells with optical microscopes, but atoms are 1000 times smaller again, 10^{-10} metres. We can imagine how small this is by realising that the relative sizes of an atom and an apple are the same as an apple and the earth! In an optical microscope the resolution is limited by the wavelength of the light that can be detected by the human eye, but even smaller objects can be studied using an electron microscope. In an electron microscope, electrons are accelerated to very high velocities and these take over the role that light plays in an optical microscope. We are already talking of replacing light, which is a wave, by an electron, which is surely a particle. This is just the first of many surprises that we shall meet in our exploration of the world of the atom. The electron microscope can be used to take pictures of objects as small as complex molecules and a field ion microscope can even produce a visual picture of individual atoms.

Part of this story concerns the proton, which is about 100 000 times smaller than the atom, with a size of 10^{-15} metres. How can we conceivably investigate this particle and its movements? When we study the detectors used in particle physics, we will find that there are several different techniques, for example the bubble chamber, which can be used to record the paths of these tiny particles. We can never observe particles directly, but we can follow their movements very precisely indeed, because when a collision produces new particles we can record every detail of their motion, and even their identity in many cases, by the trails they leave in various detectors.

This makes it much easier to appreciate what is happening when these small particles interact or decay because we can record a 'picture' of this event. The journey deep inside the atom will still require imagination but many of the important discoveries and results are recorded as spectacular pictures of the paths of these tiny particles.

2.2 Looking inside the atom

The next big step forward in our understanding of atoms occurred early in this century with the work of Lord Ernest Rutherford and colleagues. The experiments that were designed to investigate the structure of the atom have several things in common with our more recent experiments and so we will discuss these experiments in a little detail. The electron had been discovered in 1897 and was known to carry a negative charge and be much lighter than even the lightest atom. The atoms were known to be uncharged but the distribution of the electrons and the form of the balancing positive charge were not known at all. One popular idea was that the small electrons and balancing positive charge were spread evenly through the volume of an atom and not concentrated in small lumps. This was known as the 'plum-pudding' model of the atom.

In order to test this idea. Rutherford decided to fire a beam of α particles, which are heavy, positively charged particles, at atoms in a stationary target. What results were expected? If the α particle passed close to an electron in the atom it would be deflected by the force between the charges. However, the α particle has a mass many thousands of times that of the electron and is travelling very rapidly, so the expected deflection was less than one degree from a single collision. If the atom was filled with evenly spread positive charge then the positively charged α particle would be repelled by this charge. However, if the α particle entered the atom, then this force would be small, as these repulsive forces would all act in different directions. Consequently the α particles were expected to be only scattered through very small angles if this picture of the atom was correct. There were no particle accelerators at this time and so Rutherford used the naturally radioactive source of radium to produce the α particle beam in his experiments. The radioactive source was kept in a lead container which had a small hole providing a collimated beam of α particles. This was an early example of an experiment involving a beam, a target and a detector. The experiment concerning the W and Z bosons to be described in this book is a present-day version of this type of experiment, which probes even deeper into the atom.

In Rutherford's experiment, the positively charged particles were emitted with an energy of more than a million electronvolts. (The energy gained by an electron, or particle of the same charge, when accelerated through a potential difference of one volt is called an electronvolt (ev).) This high energy enabled them to penetrate very close to the positive charge in the atom before being deflected by electrical repulsion. Ideally, the experimental target would have been a single layer of atoms but in practice a thin gold foil containing several hundred layers of atoms was used. The α particles were detected after they had scattered from this target and the detector, which consisted of a zinc sulphide screen and a microscope, could be moved to different positions. When an α particle hit the screen a small flash of light was emitted. Providing observation was made in a darkened room this light could be observed through a microscope. A schematic view of the experimental arrangement is shown in figure 2.1. The experimenters made careful observations of the number of α particles that were scattered at various angles by moving the zinc sulphide screen to different positions. These experimental results were then compared to the theoretical predictions for the 'plum-pudding' model.

The vast majority of the α particles were not deflected at all, as was expected. However, the other results of the experiment were very surprising. Of the α particles that were scattered, many more were deflected through larger angles than expected. Some were even deflected through one hundred and eighty degrees! This surprise was expressed graphically by Rutherford – 'It was about as credible as if you had fired a fifteen inch shell at a piece of tissue paper and it

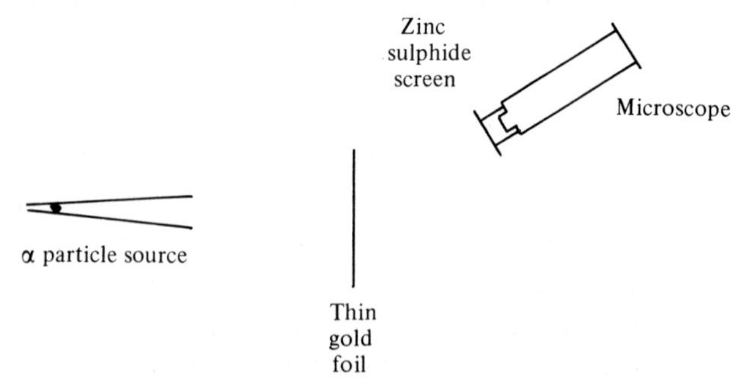

Figure 2.1 Schematic diagram of the early α particle scattering experiment which revealed that the positive charge of an atom is contained in a small nucleus. The natural radioactive source of α particles strikes a thin gold foil and the scattered α particles are detected by flashes of light on a zinc sulphide screen viewed in a darkened room with a microscope. Many more α particles were scattered through large angles than had been expected.

came back and hit you'. These experimental results could only be explained by considering the atom as being composed of a very small, heavy, positively charged nucleus, surrounded at a large distance by the much lighter electrons. In this case, the incident α particles would frequently not be deflected at all, because a large volume of the atom is completely empty. The large deflections occur when the α particle approaches very close to the concentrated positive charge of the nucleus. This positively charged nucleus is at least 10000 times smaller than the atom itself. Electrons, which are each negatively charged, move through the remaining volume of the atom, balancing the positive charge on the nucleus and keeping the atom electrically neutral. Since the electrons are also very small, only a tiny fraction of the atom's volume is occupied by matter. We have now started our journey into the atom, but to reach the smallest known pieces of matter we have much further to go. These results early in the twentieth century led directly to our current view of the atom. Rutherford had shown that the results of a scattering experiment could be used to learn about the internal structure of an atom. We shall see that similar higher-energy experiments have been used to probe the structure of the nucleus and more recently the proton.

2.3 New ideas from the quantum world

We can compare the movements of an electron around the nucleus in an atom with those of the earth orbiting the sun in our solar system and we immediately find an interesting difference. We can predict the exact position and speed (velocity) of a planet or any other object which is much larger than an atom very accurately, but this is not true for an electron inside an atom. We can either know its position or its velocity well, but never both simultaneously. At first sight this does not seem sensible but many more of the familiar properties of everyday life are changed in the atomic world. We will be considering the properties of matter at very small distances and at very high energies. We shall need to introduce the most important ideas from both relativity and quantum mechanics in order to describe particles in these extreme conditions. These two theories, which were developed earlier in this century, are not very important in everyday life when we are dealing with objects containing large numbers of atoms travelling at speeds much less than the speed of light. However, in our study of the atom these new ideas are essential.

Wave–particle duality

In everyday life there seems to be a very clear distinction between matter and waves. All matter is composed of atoms and, although small, these are particles. However, the various types of electromagnetic radiation, for example the light from the sun, are waves. The coloured spectrum which is obtained when this light is passed through a glass prism shows that these waves have many different wavelengths (the distance between successive crests of the waves).

An even more dramatic illustration that light has the properties of a wave can be seen when light is passed through a small hole. If light was carried by small particles then we would expect to see light on the screen only in the geometrical area behind the hole. In fact the pattern has bright and dark regions which extend outside the area directly behind the hole as shown in figure 2.2. Surprisingly, if the hole is made smaller, the pattern of the light and dark regions gets even larger! This interesting effect of waves is called diffraction. Now, before you try to check this for yourself, be warned that the size of the hole must be comparable to the wavelength of light before this effect can be readily observed. As this wavelength is less than one-thousandth of a millimetre, only a very small pinhole or slit will be effective. A similar effect is observed when a fine wire is illuminated by a light source, when a more complicated pattern than a geometrical shadow is produced.

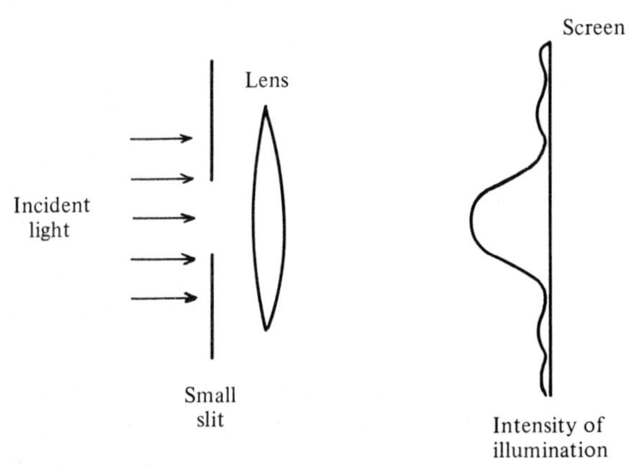

Figure 2.2. Parallel light illuminates a small slit which has a comparable size to the wavelength of the incident light. After passing through the slit the light is brought to a focus on a screen by a lens. The intensity of illumination varies across the screen, in a series of light and dark lines, caused by the diffraction of light through the slit. If the slit is made smaller the overall width of the pattern is increased.

This idea can be turned around to our advantage if we wish to study even smaller holes, for example the spaces between atoms in a crystal. In this case X rays, which are electromagnetic radiation with a much smaller wavelength than light, replace the beam of light. In a crystal the atoms are arranged in a regular lattice and so the X rays produce a diffraction pattern which can be used to work out the spacing between the atoms. We are going to show that matter can also behave like a wave and we shall use a diffraction experiment to illustrate this. The ideas of diffraction will also be used in our search for smaller objects inside the protons.

The first evidence that matter also has wave properties came from Clinton Davisson and Lester Germer's studies of electron diffraction with crystals in 1927. The electron, one of the constituents of the atom, produced a diffraction pattern and was definitely behaving as a wave. This had been predicted a few years earlier by Prince Louis de Broglie (a French aristocrat), who had even calculated that the wavelength (λ) of such particles would depend on their momentum (p) and be given by $\lambda = h/p$ (where h is Planck's constant). So as the momentum of a particle beam increases, the effective wavelength of the particle decreases. For a beam energy of 50 000 eV the equivalent wavelength is of the order of 10^{-11} metres and can be used to investigate the spacing of atoms in crystals. (To avoid writing too many numbers we usually abbreviate 1000 eV to 1 keV and 1 000 000 eV to 1 MeV). This idea that a particle exhibits wave properties, with a wavelength which gets smaller as its momentum increases, is a crucial idea to which we will refer several times.

What about waves, do they ever behave like particles? The first evidence for this came from studies of the photoelectric effect, where electrons were emitted from a metal surface which was exposed to light. In 1905, Einstein suggested that in the photoelectric effect, energy was transferred from the light to the electron in a single, very rapid interaction. The electron was not continuously absorbing light, it either received a packet of energy or nothing at all. We call this packet of energy a photon and its energy (E) depends on the wavelength (λ) by $E = hc/\lambda$ (where h is Planck's constant and c is the speed of light). The energy of the photon is also given by $E = hf$ where f is the frequency. The frequency multiplied by the wavelength equals the velocity for any wave. It takes a minimum amount of energy to eject an electron from a surface, and so this idea leads to a dramatic prediction. If the wavelength is large, the photon energy is small, and if this is too small no electrons at all will be emitted, however intense the light or however long it shines. If the wavelength is short then the energy of the photon will be large and the electron will be emitted with a predictable maximum energy. These predictions were verified in every detail by the American Robert Millikan in 1916.

We now refer to this close relationship between matter and waves as wave–particle duality. Waves and particles are interchangeable. Sometimes we need one view, at other times another, which is confusing at first sight as it seems to be imprecise. However, in the quantum world we cannot predict the outcome of each measurement; the atom is unpredictable and the familiar rules do not always apply.

An electron in an atom can have several energies, but each of these is well defined. It cannot have any energy of its choice, just a few special ones called discrete energy levels. This is completely different to our normal experience where we can drive at many different speeds and throw balls with velocities which cover a wide range. This restriction of choice is called discreteness and in an atom many quantities can only take certain discrete values.

The electrons are confined to the region of the atom by the attractive electrical force of the nucleus. Any confined system necessarily has certain preferred frequencies of vibration. Consider the motion of a plucked guitar string which is fixed at both ends. The string will resonate and continue vibrating for frequencies where only an exact number of half waves fits between the two ends. When we consider an atom we are dealing with three independent dimensions but again the same principle applies. As the electron is confined it can only exist as a stable part of the atom if it has one of these allowed wavelengths. In these cases one could imagine that an exact number of wavelengths will fit into the atom. We have already shown that the wavelength of an electron is related to its energy. Consequently the small number of allowed wavelengths means that the electrons can only exist in certain allowed energy levels.

The uncertainty principle

Consider the simultaneous measurement of the position and momentum of an electron or other particle, where the momentum of a particle is its mass multiplied by its velocity. In classical physics these measurements can be made with an accuracy limited only by the measuring equipment. In quantum mechanics we find that there is a fundamental limit, given by the so-called Heisenberg's uncertainty principle (after Werner Heisenberg, its German formulator), to this accuracy. This requires that the error in position multiplied by the error in momentum of a simultaneous measurement must always be greater than a very small constant h (Planck's constant). Beyond a certain point any improvement in accuracy on position measurement *must* be accompanied by a worsening of momentum measurement and vice versa.

Consider the observation of the position of a small object, using a microscope. Light has to be scattered from the object in order to see it and this scattering automatically disturbs it. This disturbance is increased when shorter-wavelength light is used to increase the accuracy of position measurement, because this carries more momentum. This principle is also applicable to another situation, the measurement of energy. If the time period that we specify is very short, the uncertainty principle can even temporarily affect the conservation of energy. When considering time intervals familiar to us in daily life, the conservation of energy is exactly obeyed. However, when we consider extremely short time intervals the uncertainty principle requires that there be a large uncertainty in the energy so energy conservation can be violated for short periods. The 'books' have to be balanced over longer periods but large amounts of energy can be 'borrowed' if the time interval is very short.

Quantum mechanics, which was developed over 50 years ago, replaces the predictability of classical physics with information in the form of probabilities. The location of an electron cannot be known precisely but the probability of its being in various parts of the atom is known. For this reason the expression 'electron cloud' is sometimes used. It is important to realise that this refers to the probability of finding an electron at a given position in an atom, and not its actual size. The stability of the atom with negatively charged electrons surrounding a positive nucleus cannot be understood using classical physics. The attractive force between these oppositely charged objects should lead to the collapse of each atom because the electron loses energy when it is being accelerated and should gradually move towards the nucleus. However, the uncertainty principle requires that the low-momentum electrons which are found in atoms must remain separated from the nucleus, and this guarantees the stability of the atom.

Pauli exclusion principle

Some atoms have many electrons; how are those distributed among the energy levels? We normally find that objects prefer the lowest energy state, but are the electrons all to be found in the lowest energy level? The electrons belong to a class of particles called fermions. A feature of fermions is that it is not possible for two particles with identical energies and other properties to exist in the same region. This powerful idea, originated by the Austrian Wolfgang

Pauli, is called the Pauli exclusion principle and results in the electrons in atoms occupying a number of the lowest energy levels. The detailed way in which these levels are filled governs the chemical properties of an atom, because in a chemical process only the electrons in an atom are actively involved.

The spin of a particle is one of its properties which enters into the Pauli exclusion principle. We can visualise the spin of a particle as being similar to the earth's rotation about its own axis, but on a much smaller scale! These classical descriptions of quantum mechanical effects can lead to difficulties if followed too literally. They are, however, a useful way of visualising some of the properties of these small particles. In the quantum world, spin can only have certain values, in a similar way to the discrete energy levels we met for electrons in the atom. Spin is a form of angular momentum, which is also subject to the uncertainty principle. The spin of a particle is quoted in units of Planck's constant h.

All known particles have either half integer or integer spin and are called fermions or bosons respectively. Most of the particles that we meet, including the electron, proton and neutron, have a spin of 1/2. They are called fermions and all obey the Pauli exclusion principle. However, some particles, including the pion, have a spin of zero and do not obey this principle.

If we measure a spin in the quantum world we can only determine the magnitude of the spin and its component along one direction. Furthermore, this measurable component can only yield values which differ by one unit of spin and cannot, of course, exceed the total spin. Consequently, for particles with a spin of 1/2, the only allowed values for its projection are $+1/2$ and $-1/2$. We can choose this single projection to be in any convenient direction and if it is along the particle's direction of motion, we refer to $+1/2$ as right-handed and $-1/2$ as left-handed spins respectively. A right-handed particle has its spin along its direction of motion whereas for a left-handed particle its spin points opposite to its motion.

An electron will remain in a stable energy level in an atom unless it is given some extra energy. When this extra energy is large, it may be enough to overcome the attraction of the nucleus completely, and the electron escapes from the atom, leaving a positive ion. If the energy is not large enough for this then only certain specific energies can be accepted by the electron. For an electron to be raised to a higher energy level the bundle of energy, or photon, received must be exactly matched to the extra energy needed to change to a higher energy level. Once the electron is in the higher energy level it can return to a lower level, if there is a free space, but only by emitting a photon to conserve the energy. Every change of energy in the atom involves receiving or emitting a photon. We are very familiar with these photons as they make up every light wave, radio wave, X ray or other electromagnetic wave that we encounter. The light we receive from the sun comprises enormous numbers of photons, most of them emitted from an atom in the way we have described. For our studies of atoms and smaller objects, the idea that energy is always carried as a quanta of energy is a very important one.

2.4 Structure of the nucleus

We have now discussed some of the important ideas in the quantum world, so let us return to the nucleus. So far all we know is that it is positively charged and much heavier than the electron. The same techniques that had been used to identify the presence of the nucleus were now used to probe its internal structure. In the period 1920–30, research at the Cavendish laboratory at Cambridge was dominated by the study of the nucleus.

The naturally radioactive sources of α particles were again used to supply a beam, with a zinc sulphide screen which emitted a flash of light when struck by an α particle acting as a detector. Various elements were used as targets and some were found to emit a natural penetrating radiation under such bombardment. In Paris, Irene Curie and Frederic Joliot found that this neutral radiation from a beryllium target could even eject protons from a thin absorber of paraffin wax. Shortly afterwards, in 1932, James Chadwick, one of Rutherford's colleagues, repeated this experiment with a strong radioactive source and found that this neutral radiation could even knock forward heavier nuclei, including nitrogen. These results were consistent with the idea that a neutron, a neutral particle with a similar mass to the proton, was being emitted from the beryllium nucleus. This neutron was heavy and energetic enough to dislodge another nucleus from the paraffin wax. Further experiments confirmed the existence and measured the properties of the neutron. We now know that the reaction between an α particle and a beryllium nucleus was producing a carbon nucleus and a neutron.

The nucleus of a hydrogen atom is simply a proton. The nuclei of all other elements contain both neutrons and protons. The heavier elements have more neutrons than protons in the nucleus and these constituents are packed closely together, unlike the electrons in the atom. Protons and neutrons have a similar mass, about 1800 times the mass of an electron, and are positively charged and neutral respectively.

Why are there more neutrons than protons in the nucleus of most atoms? The protons are positively charged and the electrical repulsion between them has to be balanced by an attractive force to maintain nuclear stability. The strong nuclear force, which we discuss in the next section, is generally attractive between protons and neutrons. The neutrons provide an extra attractive force and also tend to keep the positive charges of the protons more separated from each other, which reduces the electrical repulsion. For each element the nucleus contains the same number of protons as electrons, but the number of neutrons can differ. These slightly different forms of an element, which have identical chemical properties, are called isotopes. Most elements, for example carbon and oxygen, exist as different isotopes.

Most nuclei are completely stable but many others are radioactive and decay with lifetimes ranging from a fraction of a second to millions of years. The radioactive emission of α particles, which are helium nuclei consisting of two protons and two neutrons, reduces the nuclear charge and mass and changes the nucleus to one of a different element. In radioactive β decay a nucleus spontaneously emits an energetic electron. In this decay the nuclear mass is almost unchanged as the electron is much lighter than a proton or a neutron. The α and β particles typically have energies of a few million electronvolts. Sometimes a nucleus emits a gamma ray, which is electromagnetic radiation with a very short wavelength, to lose energy and in this case the element is unchanged. This is analogous to the emission of visible light or X rays when an electron drops to a lower energy level in an atom.

There are two well-known processes where large amounts of energy can be produced by converting some of the mass of the nucleus into energy.

If a neutron with a small amount of energy strikes a uranium nucleus, then the nucleus breaks into two similar sized pieces in a process known as nuclear fission. When this happens, over 100 million electronvolts of energy are released and a few other neutrons are emitted. These secondary neutrons repeat the process and a chain reaction can develop. In a nuclear bomb the volume of uranium is above a critical size so that the secondary neutrons accelerate the chain reaction before leaving the surface. In this case the very rapid conversion of nuclear mass into energy occurs for so many nuclei that it results in an explosion.

Nuclear fusion is the other method of extracting energy from the nucleus. This is the process which creates most of the energy emitted by the sun and other stars. In nuclear fusion two light nuclei interact to form a heavier nucleus, with the release of some mass energy. One example is the sequence of reactions where hydrogen is converted into helium.

2.5 The four fundamental forces

We have now identified the electrons, protons and neutrons which are inside the atoms that make up our world. Now we must start to examine the forces that act on these particles. The forces that are important in day to day activities are varied and seem very complicated. This is mainly due to the extremely large number of atoms involved in any object which is large enough for us to see. The frictional forces between the wheels of a car and the road and the force which supports us in a chair are all produced by the electromagnetic forces between atoms. When we isolate the fundamental forces between particles which give rise to these large-scale forces we find a much simpler picture. There are only four known fundamental forces and so we are able to study each of them carefully. Each force has a different strength and has a very different dependence on the separation between the interacting particles.

Gravitational force

A vital force for all of us is the gravitational force which, acting between our mass and that of the earth, is responsible for keeping us on the earth's surface. This force also attracts the earth and other planets to the sun and keeps them in stable orbits. The gravitational force decreases as the separation between particles increases, the decrease being proportional to the square of the distance. This relationship is known as an inverse square law. In the interactions of the atoms that we will be discussing, this well-known force is surprisingly unimportant. The gravitational force between two electrons is incredibly weak compared to the corresponding electric force between them, as the electrical force is around 10^{43} times more powerful! The gravitational force achieves importance in the 'real' world because it is always attractive and as there are so many atoms involved the force becomes significant. The weakness of this force beween atoms can be illustrated by considering the gravitational force between two people. Each atom has a size of about 10^{-10} metres and so occupies a volume of around 10^{-30} cubic metres. As the gravitational force between two people is negligible, and we each consist of over 10^{28} atoms, this force between single atoms must indeed be very small.

Electromagnetic force
The other fundamental force with which we have daily contact is the electromagnetic force. The combination of the electric and magnetic forces into a single electromagnetic force was a triumph for the Scottish scientist James Clerk Maxwell in the last century. The strength of the force between charges depends on their separation in the same way as the gravitational force and gets weaker when the charges are further apart. However, in this case there are positive and negative charges and unlike charges attract and like charges repel. The electric force is crucial to the existence of atoms and is very important in the interaction between elementary particles. It is only dominated by the gravitational force on the large scale because of the fact that there are equal amounts of positive and negative charges and most of the electric forces cancel each other out in large objects.

Weak nuclear force
The third fundamental force is the weak nuclear force which acts between all types of particles. This force is not well known, although it is responsible for some very well-known phenomena. It is this interaction which causes the radioactive decay of nuclei and is responsible for many of the reactions by means of which the sun and other stars produce energy. The force is not commonly experienced because it only operates over distances of less than 10^{-17} metres! It is weaker than the electromagnetic force but much stronger than the gravitational force. The lifespan of energy generation from the sun is controlled by the strength of the weak nuclear force, and so it is important for our existence that it is much weaker than the electric force! Although this force is relatively weak, it is very important for reactions where the other stronger forces cannot operate. The best example of this is the neutrino, which does not 'feel' the electromagnetic or strong nuclear forces. The neutrino can only be created and interact by the intervention of the weak force and so the neutrino is a very useful tool for studying the properties of this force. These properties turn out to be very unusual but we will return to this detail in the next chapter.

Strong nuclear force
The strong nuclear force is the strongest of all the forces but also has a very short range. This force acts over distances up to 10^{-15} metres, about the size of a proton, and is identical for both neutrons

and protons, which are both called nucleons. This generally attractive force is responsible for the long-term stability of the nucleus and prevents the repulsive electric force between protons from destroying every nucleus. This force has no effect at all over distances greater than 10^{-15} metres and so it is only experienced in the confines of the nucleus or in the interaction between elementary particles. Particles which experience the strong nuclear force are called hadrons and these include the proton, neutron and pion. Any particle which does not feel the strong nuclear force is called a lepton. The most well-known leptons are the electron, muon and neutrino.

Summary of forces

The main properties of these four forces are summarized in figure 2.3. As the gravitational force is so weak it can be neglected in all of the particle interactions to be discussed here. This leaves the weak nuclear force as the only force which interacts with *all* of the fundamental particles. This story is about the search for the carriers of this weak nuclear force.

2.6 What carries the force?

Some of the forces that we have discussed act over very long distances. The sun attracts the earth even though they are separated by millions of kilometres but how is the force transmitted from one object to the other? Similarly the mechanism for the force between two separated charges needs to be understood. In classical physics this action at a distance is usually dealt with by introducing the

COMPARISON OF FORCES

Force	Relative strength	Dependence on distance	Particles that 'feel' force
Gravitational	10^{-45}	Inverse square	All
Weak nuclear	10^{-7}	Short range	All
Electromagnetic	1/137	Inverse square	Charged
Strong nuclear	10	Short range	Hadrons

Figure 2.3. A summary of the properties of the four fundamental forces arranged in order of increasing strength. The familiar gravitational force is extraordinarily weak but it does interact with all particles at any separation. The two nuclear forces are little known because of their very short range of interactions.

concept of a field. A charged particle is considered to be surrounded by an electric field which then interacts with other charges.

However, an alternative description of the way in which the force is transmitted is provided by the uncertainty principle which as we saw earlier is an important idea in the world of the atom. This idea has some very interesting consequences in developing our idea of fields. A particle can be created for a short time because of the uncertainty principle even when no extra energy is available. However, the more energy it requires, the shorter the time it is able to exist and this fleeting particle is called a virtual particle. If the virtual particle is very heavy, then it can only exist for a short time, and if it is a force carrier then its range of interaction will be short.

These ideas are essential ingredients of one of the most precise theories in science, known as quantum electrodynamics. This theory predicts the results of experiments involving the electromagnetic interaction, with an amazing precision of one part in a million million. Every charge is continually emitting and reabsorbing virtual photons, which only exist for very short periods. When one of the virtual photons is absorbed by a different charge, the electric force has been carried between the charges. This idea of a force being transmitted by the exchange of a virtual particle is a crucial idea. It is shown for the electric force in figure 2.4*a* in the form of a diagram

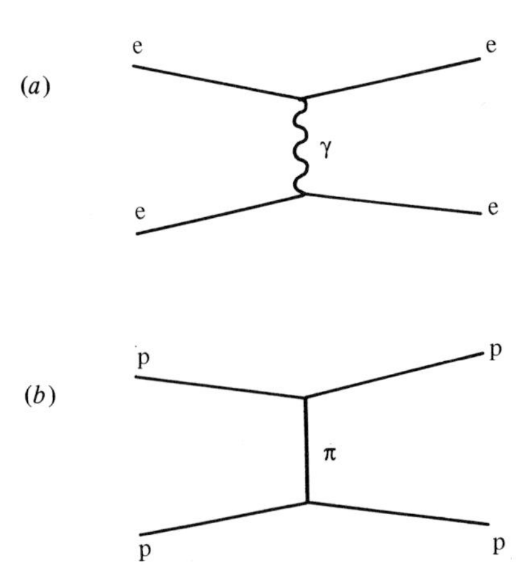

Figure 2.4. (*a*) The electromagnetic scattering of an electron by another electron where the two incoming electrons transmit this interaction by the exchange of a virtual photon. As a photon has zero mass this interaction can occur over vast distances. (*b*) The corresponding diagram for the strong nuclear force between two protons (P) which is carried by a virtual pion. The pion is a relatively heavy particle and so the range of this interaction is very short.

introduced by Richard Feynman, which is widely used for calculations of these forces.

We can estimate the maximum range of a force carried by a virtual particle of mass m in the following way. Assume that the virtual particle is only allowed to exist for a short time, Δt, and that it travels with a velocity close to c, which is the speed of light. Then the range R of this force will be given by $R = c\Delta t$. The uncertainty principle relates the uncertainty in energy ΔE to the time taken by the measurement Δt by $\Delta E \Delta t \geqslant h$. As the uncertainty in energy equals the mass energy of the virtual particle $\Delta E = mc^2$, the maximum range of the force becomes $R = h/mc$. This maximum range is clearly inversely proportional to the mass of the exchanged particle which means that a heavy force carrier will only be able to transmit a force over a small distance.

All forces are considered to be carried by a virtual particle but each force has a carrier of different mass which gives rise to the differences in range of these forces. The electric force can be carried over very large distances because its carrier, the photon, is massless. Consider the strong nuclear force between proton and neutron, which only has a range of 10^{-15} metres. This force has a short range because the carriers of the strong force are virtual pions and these are particles which have several hundred times the mass of an electron (figure 2.4*b*). The uncertainty principle will only allow large energies to be borrowed for very short times. For the strong nuclear force the energy loan is large, and as no particle can travel faster than the speed of light, the range of the virtual pion has to be small.

In early studies of the weak interaction it was thought that this interaction had no range at all and that the particles only interacted if they were at the same point. Now we know that this force has a range of 10^{-17} metres and to explain this the virtual particles which carry the weak nuclear force must be much more massive than the carrier of the strong nuclear force, the pion. This time the virtual carriers are the massive W bosons which have masses around 100 times that of a proton. This explains the extremely short range and consequent weakness of the weak force because it can only operate over very small distances.

The gravitational force is very weak on the atomic scale. Although this force is thought to be carried by massless gravitons, there is still no firm evidence for their existence. The relative weakness of this force makes the experimental searches extremely difficult, although several searches for gravity waves have been conducted.

Finally, let us consider the mass of the virtual particles that carry the various forces. In the quantum world the uncertainty principle allows energy conservation to be briefly violated and 'borrowed' energy is continually used to create these virtual particles. These fleeting virtual particles have many properties in common with real particles, but their masses are different. Light is an electromagnetic wave, composed of real photons which each have a mass of precisely zero. The virtual photons which carry the electromagnetic force have non-zero masses!

The real W bosons are very heavy and we will be searching for these later in this story. However, the weak nuclear force is continually acting all around us, in radioactive decay and the nuclear reactions in the sun and stars. These reactions proceed by the exchange of virtual W bosons, which have much smaller masses than real W bosons, as allowed by the uncertainty principle. Our world is full of virtual W bosons but this story is about an experiment to provide the enormous energy to create the first real W bosons.

2.7 Summary

We have seen that all matter is made up of atoms, which each contain protons, neutrons and electrons. These protons and neutrons provide the mass of the atom and are located in a small nucleus at the centre of the atom. We have also introduced some of the interesting new physical properties that we find in the quantum world. Particles and waves become interchangeable and we find discrete electron energy levels in the atom. The uncertainty principle, which is a consequence of wave–particle duality, makes exact measurements impossible but it allows virtual particles to exist for short periods and to act as force carriers.

The wide range of forces that we normally encounter has been reduced to four that we believe are fundamental. The extremely varied properties and strengths of these forces can be described by introducing the concept of virtual, or short-lived, particles which carry these forces. The electromagnetic force is carried by virtual photons. When light is transmitted it is made up of a large number of real photons and so both virtual and real photons are important in our normal life. The weak nuclear force, which is responsible for radioactivity and the energy cycle of the sun, is believed to be carried by virtual W bosons. The real W bosons are predicted to be very heavy, around 100 times the mass of the proton, and until recently no accelerator had produced collisions energetic enough to create these particles.

In chapter 4, we will introduce some of the important discoveries and ideas that led to the unification of the weak nuclear and electromagnetic forces into the electroweak theory. This introduces a massive neutral Z boson and predicts the masses and other properties of both W and Z bosons very precisely. It also predicts that the Z boson and the photon are very closely related; so closely that the Z boson is sometimes called heavy light! However, before we can look more closely into the forces of nature we need to probe even deeper into the atom and look for the structure inside the proton.

Chapter Three

Quarks and leptons

3.1 Introduction

We will now investigate whether electrons, protons and neutrons are the building blocks for every single atom. As higher-beam energies become available, the structure of matter can be studied down to even smaller distances inside these already tiny particles. However, it is not inevitable that further structure is found. Take the example of the electron which was known in the last century. After years of intensive study, there is still no evidence at all for any structure within the electron. However, in the same period the nucleus has been shown to be made of protons and neutrons. This chapter describes the experiments which showed that these protons and neutrons are in turn made up of quarks. Present results suggest that the electron and other leptons, which do not feel the strong nuclear force, have no inner structure. There may be even deeper levels of structure, but only time will tell.

3.2 The electron and other leptons

The electron is a crucial part of every atom. Although its mass is insignificant its electric charge is responsible for all of the chemical reactions between different substances. The electron is measured to be smaller than 10^{-17} metres in size and for this reason we refer to it as a 'point-like' particle. We can calculate the properties of electromagnetic interactions very accurately and this has meant that the electron is frequently used to study the structure of other objects including the proton. The electron is the only lepton which exists in the atoms of our natural world.

Where do we find the other leptons? There are many other

interesting and unseen features to our everyday life. The surface of the earth is illuminated by the light from the sun and other stars. In addition, cosmic rays, which consist of many types of particles originating from the solar system and beyond, also bombard the earth. Many of these interact with atoms in the upper atmosphere and produce other particles which arrive at the surface of the earth. As a result we are traversed by a few of these particles every second.

In the 1930s cosmic rays revealed the first of many surprises about the number of elementary particles. Cosmic rays include muons (μ) which are charged, 200 times as heavy as an electron and highly penetrating. Muons do not appear to have any internal structure, and they are sometimes called heavy electrons. The muon has some different properties from the electron, but superficially it seems like an unnecessary duplication of the electron. Although the muon is heavier than an electron it has never been observed to decay into an electron and a photon. This emphasises the fact that leptons, while similar in many properties, do carry one different property. The 'electroness' and 'muoness' cannot be given up in any type of interaction. For this reason we say that an electron has an electron lepton number of one and that a muon has a muon lepton number of one. These separate lepton numbers seem to be conserved in all interactions.

The muon and electron are both charged leptons. Are there many more? Only one other charged lepton has been discovered, known as the tau (τ) and this has a mass twice as large as a proton. This was discovered in the late 1970s and again it has a separate property which forbids its direct decay to muons or electrons, unless other particles are produced to conserve the tau lepton number. The electron, muon and tau are all negatively charged, point-like particles and appear to have no internal structure. They are called leptons because they do not feel the strong nuclear force.

The only other known lepton is the neutrino (ν). These are also present in cosmic rays and are also emitted in some radioactive decays. This is a most unusual particle which is neutral, seems to have no mass and which travels with the speed of light (3×10^8 metres per second). It is so weakly interacting that most of the neutrinos which strike the earth pass through it without interacting at all! The bizarre properties of the neutrino meant that the early predictions of its existence were not universally accepted. However, this particle is now routinely used in particle physics experiments and its properties have been studied in great detail.

There are several types of neutrino, each associated with a charged lepton. The neutrinos of the electron (ν_e) and muon (ν_μ) type are well established and it is believed that a ν_τ also exists. These are distinct particles and have not been observed to convert from one to the other. The neutrino has no charge and so it only interacts by the weak interaction. The concept of a particle with no mass is certainly a little difficult to comprehend. We have already seen that mass and energy are interchangeable and the neutrino is just an extreme case of this with all of its energy committed to its motion. The list of known leptons is now complete. There are three charged leptons each with an uncharged massless neutrino of the same type.

The heavier leptons can decay into the lighter ones but only if neutrinos are also produced to enable the separate lepton numbers to be conserved. The muon (μ^-) decays to an electron (e^-), neutrino and antineutrino: $\mu^- \rightarrow e^- + \nu_\mu + \bar{\nu}_e$. The bar over the second neutrino indicates an antineutrino which is needed to cancel the electron's lepton number. Another antiparticle, the antiproton, is a key element in our story, and so we must now investigate antiparticles.

3.3 Antiparticles

In 1928 the English physicist Paul Dirac published an important paper which was the first to incorporate relativity into the quantum theory of the electron. This contained an equation which had solutions which at first sight corresponded to negative-energy electrons! The equation was correct and indeed it could be interpreted as predicting the existence of positive electrons, later called positrons. This antiparticle of the electron has the same mass as the electron but the opposite charge and in 1932 the first positron was identified by Carl Anderson in a cosmic ray experiment. It is now well known that an energetic photon can convert all of its energy into the creation of an electron–positron pair when passing through matter. This spectacular evidence of energy being converted into matter can be seen in many of the energetic interactions that we will soon be studying. In pair production the two created particles share the surplus energy.

The antiproton is a negatively charged version of the proton. It has the same mass as the proton and so requires a high-energy collision for its production. The antiproton was discovered in 1955 in an experiment studying proton–proton collisions with a beam momentum of 6 GeV. (In particle physics, where we are studying the structure of the proton, energies exceeding 1000 million electronvolts (GeV) are involved and this is the unit we shall generally be using.) A proton carries a property which we represent as baryon number, which is conserved in all interactions. The systematic observation of many reactions leads to the assignment of a negative baryon number to the antiproton. As the antiproton has a negative baryon number, it can only be produced in association with another baryon otherwise the conservation of baryons would be violated. Since these early discoveries, an antiparticle has been discovered for every particle. So the list of fundamental particles had to be doubled at a stroke! Each antiparticle has the same mass but the opposite electric charge as well as opposite lepton and baryon number. Antiprotons are crucial ingredients in this story of the search for the W and Z bosons. When antimatter collides with matter, the mass energy of these two particles is converted into energy, which is used in the creation of new particles. We will see later that it was necessary to produce specially the antimatter for this experiment and also to keep it safe from colliding with matter! Whenever energy is used for the creation of particles, equal quantities of matter and antimatter are always produced. Yet our part of the universe seems to consist purely of matter. This is a mystery to which we will return in a later chapter.

3.4 Investigating the proton – more new particles

How can we tell if the proton has even smaller particles inside it? The natural beams of α particles were very useful in identifying the nucleus, but they do not have sufficient energy to probe the inside of the proton. We have already seen that effective wavelength (λ) of a beam is related to its momentum (p) by $\lambda = h/p$. We need very high-momentum beams to make the wavelength small enough to investigate the inner structure of the proton. During the 1950s and 1960s more powerful accelerators were developed which at last had enough energy to allow a search for structure within the proton. The initial experiments involved colliding energetic beams of protons against a hydrogen target. As the nucleus of a hydrogen atom is a single proton these were the first proton–proton collisions to be studied in an accelerator. Later we shall see that there are less complicated ways of searching for structure but these first attempts observed the reaction products of the collision. It was hoped that by studying these fragments it would be possible to deduce what was inside the proton, if indeed it did have an inner structure.

One of the most powerful types of detector for identifying these new particles was the bubble chamber. In this device a large tank of liquid hydrogen is kept at a very low temperature of around -250 degrees centigrade with the liquid maintained at a high pressure. If a beam of protons enters the chamber some of these will interact with

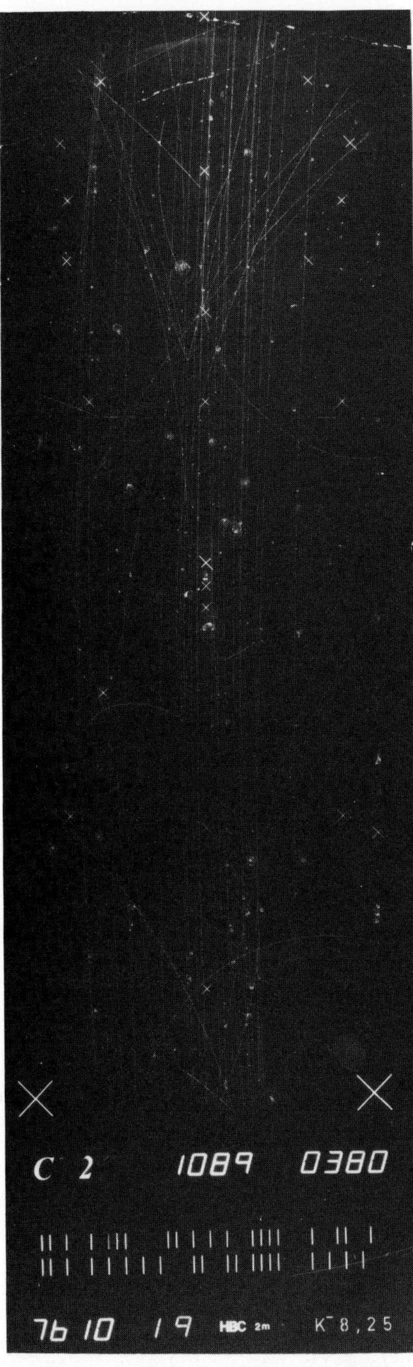

Figure 3.1. A hydrogen bubble chamber photograph which records the passage of a high-energy beam of negatively charged kaons through its volume. The path of each charged particle is recorded by small bubbles and these paths are curved because the chamber is in a region of high magnetic

the protons of the liquid hydrogen. The beam particles and the charged particles produced in the collision pass through the chamber, causing ionisation of hydrogen atoms. If the pressure is suddenly reduced, the hydrogen starts to boil, and this occurs preferentially along these paths of ionisation. By illuminating a flash system, the 'bubbles' can be photographed by several cameras, after a delay to allow the bubbles to grow to a convenient size. The bubble chamber film then retains a record of the collisions and the paths of all charged particles produced in them (figure 3.1). Usually the chamber is placed in a region of large magnetic field which bends the charged particles in an arc of a circle. The radius of the circle increases with the momentum of the particle and also allows the negatively and positively charged particles to be distinguished. Unfortunately, uncharged particles like the neutron are not detected by a bubble chamber.

No obvious pieces of the proton were produced in these collisions; in fact the protons usually stayed intact. However, other particles were produced with masses and properties which had never been seen before. Although the search had been aimed at detecting simple objects inside the proton, the observations became more and more complex. This new information later proved invaluable and once the data had been collected, great ingenuity was used to detect a simple underlying structure to explain the many new particles that were created in these energetic collisions. Many new conservation laws which seemed to govern the allowed products of such collisions were also discovered in these experiments. These were gradually revealed by studying the collision products in many different reactions. These observations were consistent with the fact that particles carried other properties than electric charge, for example baryon number and lepton number, which were conserved in all interactions. However, other new properties, for example strangeness, were only conserved in certain reactions. Many of these new particles decayed after 10^{-8} seconds but as they were produced travelling at very high speeds their decays could be observed directly. These new particles included the pion, kaon and lambda particles, all with different masses and decay properties. Many other particles decayed after the even shorter time

Caption for fig. 3.1 (*cont.*)
field. Most of the incoming beam particles pass straight through the 2-metre chamber without interacting with any protons. However, several kaons do interact, some of their energy is converted into mass and new particles are created. The small spirals correspond to electrons which have been ejected from the atom by the incident particle.

of 10^{-23} seconds. No trace of these particles could be identified directly in any detector, so how were they discovered?

If very short-lived particles are travelling with the speed of light, 3×10^8 metres per second, they can only travel around 10^{-14} metres before they decay, even allowing for the effect of relativity on the particle lifetime. This is an incredibly small distance, similar to the size of a nucleus, and yet remarkably such particles can be identified.

Consider a collision between two protons at very high energy where sometimes the energy of the collision can produce an excited state of the proton. This excited state can decay by the strong nuclear interaction in 10^{-23} seconds into a proton and a pion. There is one very important equation from relativity that we will need in this discussion. Let us define E as the total energy, P as the momentum and M as the mass of a particle; then $E^2 = P^2 + M^2$. The units have been chosen so that the velocity of light and Planck's constant both equal one. This means that energy, momentum and mass have the same units and we will frequently use electronvolts for this.

Radioactive decay occurs as a result of changes in the nucleus and a typical energy for β particles is a million electronvolts (MeV). The approximate rest masses of the electron and proton are 0.5 MeV and 1 GeV (1000 million electronvolts) respectively in these units. Notice that we refer to the mass as the rest mass because when particles travel rapidly their masses increase. All reference to mass will refer to the particle at rest unless otherwise specified. The energy equation we have been discussing applies equally well to a single particle or a system of particles. Rearranging the equation we find that the mass (M) is given by $M^2 = E^2 - P^2$. For a single proton if we substitute its observed total energy (E) and its momentum (P) we will find its well-known mass of around 1 Gev.

Consider an interaction between a moving proton and a stationary proton as shown on figure 3.2. In this collision the beam energy was converted into mass energy to create an extra positively charged and negatively charged pion (π^+, π^-) and the kinetic energy of the particles. We can measure the momentum of each particle from its curvature in a magnetic field but we need to know the particle mass in order to deduce its energy. The identity of a particle can sometimes be deduced from the number of bubbles per unit length or other property of the bubble chamber track. However, frequently the mass of a particle has to be guessed and then checked by using the conservation of energy and momentum. If the particles are assigned the wrong identity then the energy and momentum conservation laws will not be obeyed. As this procedure of guessing then checking is

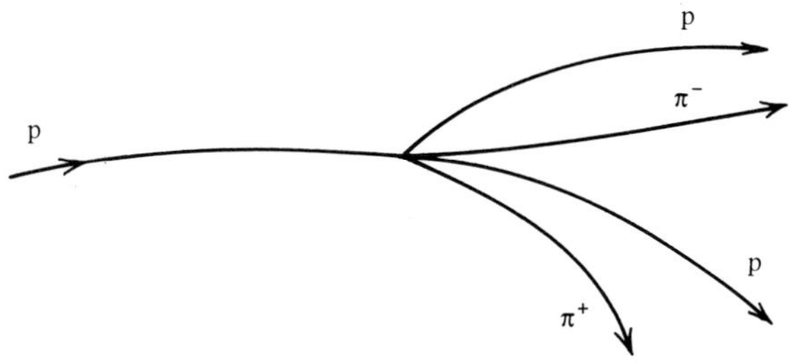

Figure 3.2. A schematic view of the tracks observed in a hydrogen bubble chamber recording a high-energy proton–proton collision. Some of the collision energy has been converted into matter producing an extra positive and negative pion. The chamber only detects moving charged particles and so the target proton is unseen. The momentum of a particle can be measured from the curvature of its track in the known magnetic field. Sometimes a short-lived particle, which decays to the final-state particles, is produced but this can only be deduced from measurements of the observed tracks and is not seen directly.

very tedious it is performed by using computers which then identify the possible outgoing particles.

We have no immediate way of knowing whether any of the final four particles in figure 3.2 resulted from the decay of a very short-lived parent. Let us investigate this possibility for the downward moving proton and positive pion. If these two particles resulted from the decay of a single parent, then energy and momentum conservation means that their combined energy and momentum is equal to that of the parent. We must remember when combining these quantities that for energy we can just add the magnitudes and that for momentum we need to allow for the directions and add as vectors. This is because energy has only a magnitude and we call this a scalar quantity. On the other hand, momentum has a magnitude and a direction and we refer to this as a vector quantity. When we have done this we can use the equation to calculate the mass of the parent particle.

In one event this has little significance but if many thousands of such events are studied we can then record the resulting effective mass values of all events (figure 3.3). Sometimes the proton and positive pion seem to be produced independently but the peak shows that most of the time they result from a massive parent which is slightly heavier than a proton. In fact the parent has a mass which is around 1200 MeV. The peak in the histogram is quite broad and yet if we measured the proton mass we would always get a very similar answer. Clearly we have to allow for measurement errors which will tend to broaden the peak but the most important reason for the

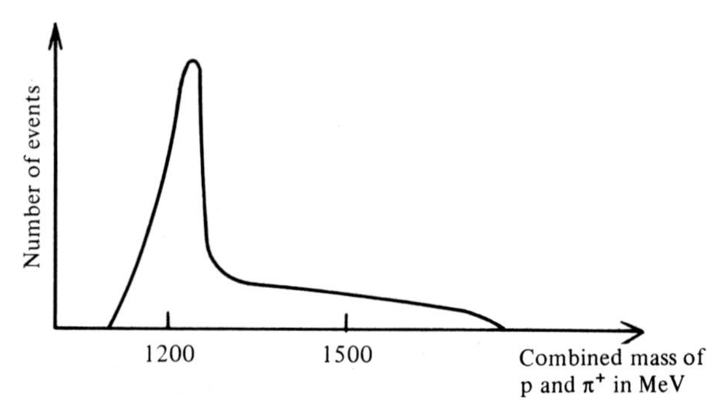

Figure 3.3. A typical combined mass distribution from a large sample of collisions of the type shown in figure 3.2. In each such event the momentum (P) and energy (E) of the proton and positive pion track have been combined and the combined mass (M) calculated from $M = \sqrt{E^2 - P^2}$. The clear peak in this distribution shows that in many of these events a short-lived particle with a mass near 1200 MeV was produced. This particle, which has a positive charge of $+2$ is known as the $\Delta(1236)$, where the number represents its mass in MeV.

broadness of the peak is much more interesting and it is due to the uncertainty principle. We have seen that in quantum mechanics we cannot measure both momentum and position or energy and time to any arbitrary accuracy. If one measurement of the pair is accurate the other is forced to be poorly determined. In this case the short-lived particle only exists for about 10^{-23} seconds (Δt) and so the uncertainty on its energy (ΔE) is necessarily large because the uncertainty principle states that

$$\Delta E \Delta t \geqslant h.$$

As Planck's constant, h, is 6.6×10^{-22} MeV seconds, we find for a particle with a lifetime as short as 10^{-23} seconds that the minimum uncertainty in the energy is greater than 66 MeV. The particle lives for such a short time that this spread on its mass value or energy is a direct result of the uncertainty principle.

This technique was very successful and during a period of many years over 100 new particles were discovered with a wide range of masses and other properties. However, this posed a new problem: how could this very large number of particles and resonances be fundamental? Were there any patterns in the groups of particles that had been found? Some of the particles existed in several charged states with similar mass, e.g. the pions π^+, π^- and π^0 and the delta resonances with Δ^{++}, Δ^+, Δ^0 and Δ^-. Other particles existed only as uncharged or neutral states like the Λ^0 and ω^0.

Some of the particles that were found were called strange particles

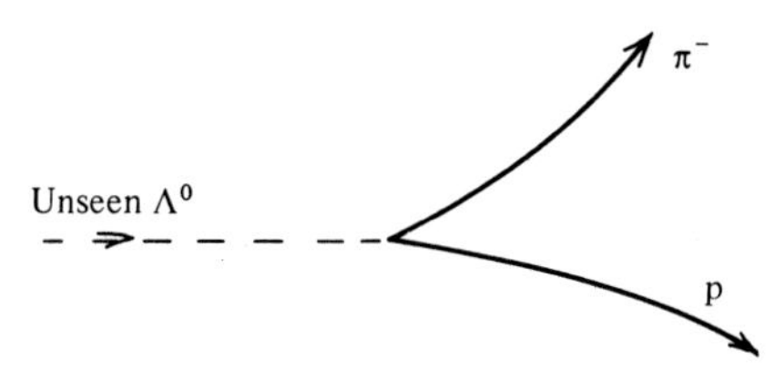

Figure 3.4. A bubble chamber can only detect charged particles and so neutral particles can only be detected if they decay to charged particles. The neutral lambda decays to a proton and a negative pion by the weak nuclear force and when it decays it leaves a characteristic V shape in a bubble chamber photograph.

because they were frequently produced but decayed relatively slowly and left visible evidence of their presence. These included the neutral kaons and lambdas, which both decay to charged particles and so leave characteristic V-like shapes on the bubble chamber film when they decay (figure 3.4). The strange particles were produced by the strong nuclear force which always conserves strangeness, a property carried by these particles and called strangeness because of the strange, inexplicably long lifetimes. These particles are produced in pairs so that the positive and negative strangeness cancels out. The K^0 is the lightest strange particle and so when it tries to decay it has to 'lose' its strangeness. No strong interaction decay is then possible and so the decay is mediated by the weak nuclear interaction. This takes much longer and allows time for the particle to travel a measurable distance. The Λ^0 is the lightest strange baryon and so its decay also cannot proceed by the strong nuclear force. From careful measurements of the decay properties of these particles it was deduced that the newly discovered particles had many different spins. Measuring this spin in units of h (Planck's constant), many had spins of 1/2, 3/2, 5/2 or 7/2 (half-integer spins) and are known as fermions. Others had spins of 0, 1, 2 (integer spins) and are known as bosons. The most important difference between fermions and bosons is that only fermions obey the Pauli exclusion principle (see p. 11).

3.5 Hadrons are made of quarks

So many new particles were found that they were clearly not elementary and so attempts were made to classify them. It was an awesome task, rather similar to the classification of elements into the periodic table, with new, short-lived particles being discovered with great regularity. These particles all interact with the strong nuclear force and are called hadrons. Some of these have half-integer spins like the proton and neutron and are known as baryons. Other hadrons which have integer spins are called mesons and these include the pion and the kaon. Gradually new ideas were developed and the most successful scheme of classification was called the eightfold way. In this scheme all baryons were constructed of three smaller objects called quarks and the mesons from a single quark–antiquark pair. All known hadrons could be described in this way by introducing just three types of quarks. This very successful idea was introduced well before there was any other evidence for smaller constituents inside the hadrons. Consequently, at the time many people believed that the quarks were just a convenient mathematical idea, which did not necessarily imply that there were physical quarks inside these particles.

When the known particles were divided into categories of different spin, these families were indeed of the same size and content as predicted by this theory. The proton and neutron were members of a group of eight particles as predicted but another theoretical group of ten particles had an interesting missing entry. The properties of this particle, known as the Ω^-, could be predicted from the other members of its group. It should have a mass of 1650 MeV and a strangeness of three and searches were made for this particle in bubble chamber pictures all over the world. In 1964, at Brookhaven National Laboratory in the USA, the first Ω^- was identified and its decay sequence, which involves many steps, is shown in figure 3.5. This was a great success for the quark model to which M. Gell-Man, Y. Ne'eman and G. Zweig had made important contributions. This

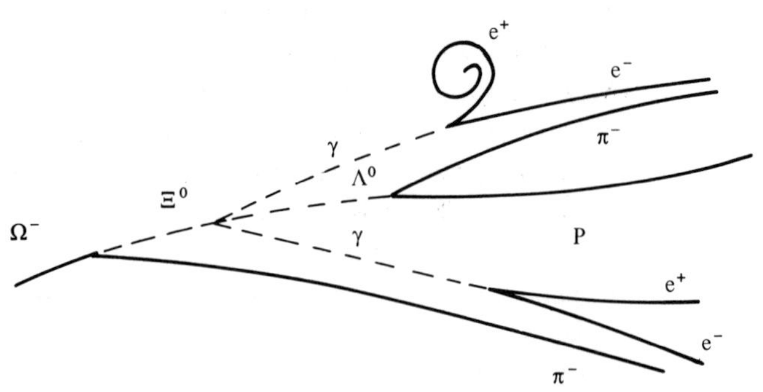

Figure 3.5. A schematic view of the decay of the first Ω^- detected in a bubble chamber. The Ω^- decays to a negative pion and a neutral Ξ^0 which after a few centimetres, decays into a neutral lambda and a neutral pion. The neutral pion decays promptly into two separate photons which each convert into a visible electron–positron pair. Finally the neutral lambda decays into a proton and negative pion. This event was remarkable in that each neutral particle could be identified uniquely.

classification scheme is still very useful today and there is still no clear evidence for any particles with properties which are not expected from this scheme.

To summarise, all the hadrons, which are particles which feel the strong interaction, can be built up in just two separate ways. Particles like the proton and neutron are called baryons and are made of three quarks. These all have half-integer spins and only occur in families of eight or ten particles each with a common spin. The only other types of hadrons, such as the pion and kaon, are called mesons and can be constructed from a quark–antiquark pair. These all have integer spins and can be arranged into families of eight particles each with a common spin. Any successful theory of interquark forces clearly has to explain why all hadrons are composed of three quarks or a quark–antiquark pair.

In the early 1960s these ideas were commonly accepted but it was still possible that this scheme was only a mathematical convenience. No single quarks had been observed, even though the proton had been subjected to very energetic collisions. The proton is very small, about 10^{-15} metres, could it really have any internal structure?

3.6 Quarks detected inside the proton

In the early 1960s a different approach was tried which yielded independent information about the structure of the proton. Near Stanford University in California there is a two-mile-long linear electron accelerator which accelerates electrons up to very high energies of many GeV. It is much easier to deduce information on the structure of the proton from a collision with an electron rather than with another proton. The electron does not appear to have any internal structure and when it approaches a proton it is deflected by the electric forces. We can deduce the internal properties of the proton by studying the way in which this scattering depends on beam energy and scattering angle.

When we discussed diffraction we found that smaller objects produced wider diffraction patterns or larger scattering angles. We also know that the higher the energy of the beam, the shorter its equivalent wavelength, and so we can use it to probe smaller distances. When an electron interacts with a proton it is deflected and emits a virtual photon which carries the electromagnetic force between the particles (figure 3.6). The energy and momentum of the incident and scattered electrons can be measured and so the energy and momentum of the virtual photon can be deduced from energy and momentum conservation. This virtual photon is the effective

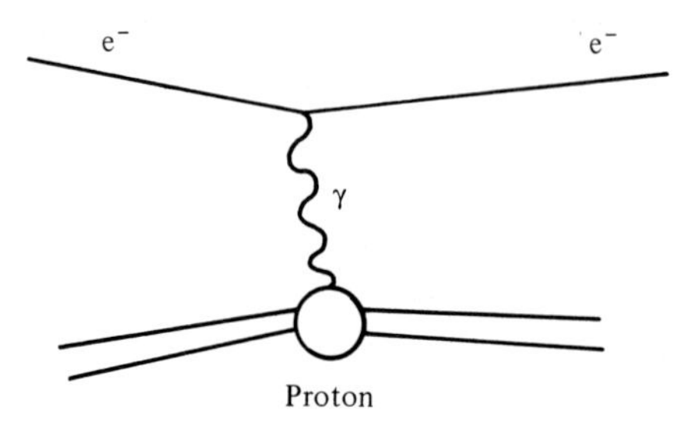

Figure 3.6. An illustration of the interaction between an electron and a proton which is carried by a virtual photon. The wavelength of the virtual photon depends on the difference in energy and direction between the incoming and outgoing electron. The most violent collision provides the shortest wavelength photon which can be used to search for small objects inside the proton.

beam which probes the proton in such an experiment. Consider the simple case where an electron beam retains its full energy but is deflected through a scattering angle, θ. If this scattering angle is increased, then the virtual photon carries an increased momentum, and as it has a smaller wavelength it is able to probe smaller distances inside the proton. The scattering angle of the electron can be related to the similar angle of the light forming a diffraction pattern. A 'point' object produces a wide diffraction pattern and this in turn would be expected to produce a similar amount of scattering at both small and large angles. An extended object has a narrow diffraction pattern and this would be expected to produce scattering predominantly at small angles.

When the first electron–proton scattering experiments were started it was thought that the proton was a 'point' charge. However, when the first results were analysed they were a major surprise as the scattering was observed mainly at small angles. The proton, although small, was not a point and for the first time its charge was measured to be spread over a distance of 10^{-15} metres. These scattering experiments continued and careful analysis of the results revealed many detailed properties of the structure of the proton.

A few years later even higher-energy electron beams were used in a similar series of experiments. The experimental detector was arranged to record even more violent collisions, known as deep inelastic scattering. In these very energetic collisions the electrons gave up energy and momentum to the virtual photon and the scattering was found to occur equally over a very wide range of angles. This was another major surprise as it indicated that the scattering was occurring off very small 'point-like' objects deep within the proton. These objects, sometimes called partons, were at least 100 times smaller than a proton and were now being 'seen' by the high-energy beam. These were quarks inside the proton but it took a great deal of experimental and theoretical work to prove it.

These scattering experiments have now been extended to very high energies with beams of electrons, muons and neutrinos. The results are conclusive – these scattering experiments reveal three valence quarks deep inside both the proton and the neutron. The spin of these quarks has been measured to be 1/2, the same as the electron, and even the quark charges can be deduced to be just fractional multiples of the electron charge e. These quarks are known as the up (u) quark, charge $+2/3$, and down (d) quark, charge $-1/3$, and the combinations uud and udd make a proton and neutron respectively. Careful studies of these scattering experiments have revealed even more details of the inner structure of the proton. When the total momentum carried by the quarks was measured it was discovered that it only corresponded to one-half of the proton's momentum. We now know the reason – because the other half is carried by particles called gluons which carry the force between the quarks and prevent any single quark from being released. As the gluons are neutral they are not detected directly by the photon, which only interacts with charged particles.

Later studies showed that there are even some antiquarks inside the proton but these carry only a small fraction of its momentum. A quark–antiquark pair can be produced by a single gluon in a similar way to the creation of an electron–positron pair from a photon. So our current picture of the proton is that there are always three more quarks than antiquarks, called valence quarks, which carry about half of its momentum. In addition there is a large number of gluons and quark–antiquark pairs. In fact the proton is a very busy place.

3.7 Electron-positron annihilation

We have now collected evidence for the existence of quarks from two independent sources, particle classification and scattering experiments, and there is a further piece of evidence for their existence. This comes from energetic collisions of electrons and positrons which annihilate and convert all their mass into energy. When this collision occurs, occasionally an electron and a positron are recreated from the energy but it is much more common to see just two 'jets' of particles moving away from the interaction point (figure 3.7). These jets usually contain several charged and neutral particles and the two

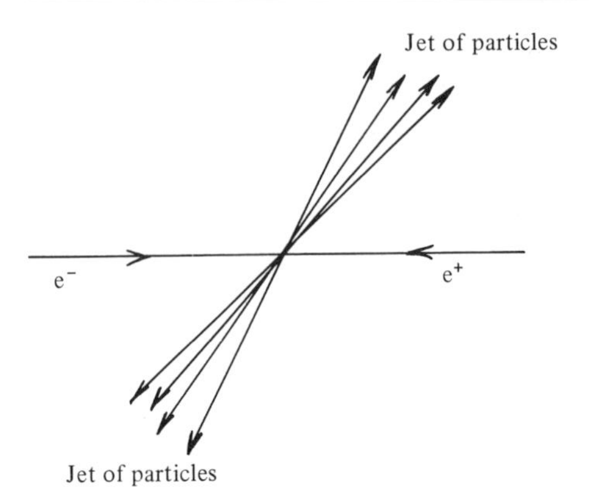

Figure 3.7. When an electron collides with a positron the mass and energy of these beam particles is converted into the energy of a single 'massive' virtual photon. This converts in a very short time to a pair of oppositely charged particles which are produced back to back with the same momentum. Sometimes this is a quark–antiquark pair and in this case two 'jets' of localised energy are produced. A jet is the observable result of the original quark after it has converted into hadrons. As the quark moves away from the antiquark, further quark–antiquark pairs are created to 'dress' the quark into a colourless group of hadrons.

systems are emitted 'back to back' in the detector with the same amount of momentum. The jets include protons, pions and other hadrons. What has this to do with quarks?

When the electron and positron collide a virtual photon is created with all the energy of the annihilation. This photon has no momentum because it is produced in a head-on collision. From the earlier equation involving energy, momentum and mass, we can see that if the momentum is zero then the mass of the photon must equal its energy. This mass is not zero and so it is a virtual photon, which exists for only a small fraction of a second, and then gives its energy to a charged particle–antiparticle pair. The electron is the lightest known charged particle and we already have discussed real photons converting into electron–positron pairs (p. 24). However, as these virtual photons have very high energy, quark–antiquark pairs can also be produced. As the virtual photon has no momentum, the quark and antiquark are emitted back to back with identical momentum. Are we finally going to see these elusive quarks? Unfortunately not, but to see why we must discuss the gluon and the property know as colour charge on the quarks.

The gluons play a crucial role in keeping the quarks together inside the proton. The interquark force is carried by massless gluons and these are the objects which hold the quarks together even in the highest-energy collision. The gluon has some similarities and some differences to the photon. The photon interacts with the charge of any particle whereas the gluon interacts with a property called colour charge which is only carried by quarks. The photon is itself neutral and so can never change the charge of a particle or interact with itself. A gluon carries colour and so can change the colour of a quark and can also interact with another gluon.

The strength of the electromagnetic force, which is carried by the photon, falls off with increasing separation between the charges. The strength of the colour force between quarks, which is carried by gluons, varies in a very different way with the separation between the quarks. At small separations the force is very weak, so weak that the quarks appear to be almost free, when they are struck in scattering experiments. As the separation increases the force becomes progressively stronger and it is thought that this prevents a single quark from escaping alone. There is a successful theory of quark interactions called quantum chromodynamics (QCD) which has many similarities to the theory of quantum electrodynamics. However, as the force is carried by coloured gluons, which can interact with each other, it is more complicated than the interaction carried by neutral photons.

What happens when a quark and antiquark are produced in an electron–positron collision at high energy? As the quark and antiquark move away from each other, the colour force between them gets stronger and stronger, until further pairs of quarks and antiquarks are produced. This is similar to the production of new poles on a magnet when it is broken into pieces. These extra quarks and antiquarks group with the original quark and antiquark pair to make particles which can exist alone like pions and protons. This process where a single quark is modified by the strong force to be suitable for the outside world is often called the 'dressing' of the quark. The original quark and antiquark end up as hadrons, and the energetic parent quark reveals itself by causing the produced hadrons to travel close together in a 'jet' of nearby particles.

So the clear observation of jets in electron–positron collisions is yet further evidence for the existence of quarks. Can we deduce how many different types of colour charges exist? There are again several different ways to do this and the clear result in each case is that there are only three types of colour charge. These different colour charges are sometimes called red, green and blue but they do not have any connection with the colours of the rainbow! There is only one similarity and that is the way that white light can be obtained from a mixture of different colours. There are only two simple ways to obtain a colourless group of quarks: a quark–antiquark pair with opposite colours or a set of three quarks each with a different colour. But these are how the mesons and the baryons are all constructed! It seems that only 'colourless' states can exist in the real world.

The best way to measure the number of colours is again in electron–positron collisions. When the virtual photon is created it produces only charged objects and it does this at a rate that just depends on the charge of the particles being produced. This is because the strength of the electromagnetic force is proportional to the electric charge of the particles involved in the interaction. The only fundamental particles that can be produced are quarks or leptons. The production of a muon–antimuon pair and an up quark–antiquark pair are shown in figure 3.8. The only difference in the processes are the charges, which are 1 and 2/3 respectively, in units of the electron charge. If we measure all the events produced by each reaction, the ratio should just depend on the charge squared and the rates should be in the ratio $(2/3)^2:(1)^2$ which is $4/9:1$. Whilst

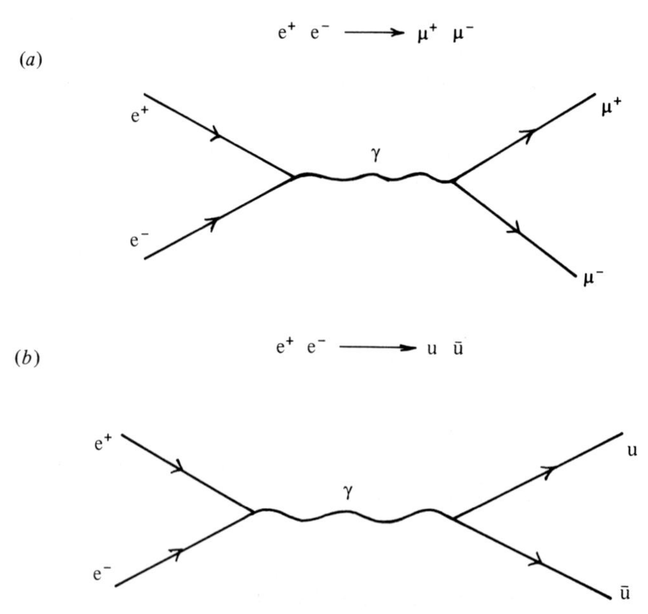

Figure 3.8. (*a*) A schematic view of an electron–positron collision, producing a massive virtual photon which then converts into an oppositely charged pair of muons. (*b*) A similar diagram for the production of an up quark–antiquark pair from an electron–positron collision. At current energies, the quarks and leptons appear to be structureless and so the relative rate of reactions (*a*) and (*b*) just depends on the charge of the up quark (u), as the muon has a charge of one. Although the up quark has a smaller charge of $+2/3$ it exists in three different colours and so it will be produced $3\times(2/3)^2 = 12/9$ times as often as the muon.

we can identify the muon pair, we cannot directly define the events containing a pair of up quarks. However, we can treat all the events which include hadrons as being produced from pairs of quarks. There are other quarks to consider, for example the down (d) quark which has a charge of $-1/3$ which will add another $(-1/3)^2$ to the quark reactions. We have already discussed strange particles which are built from quarks just like the other hadrons but each of these contains a strange (s) quark which also has a charge of $-1/3$.

So there are three quarks (u, d and s) that we have identified and we describe these as three different flavours of quarks. We can now take all these quarks and work out how often in electron–positron collisions that we should observe hadronic events compared to muon–antimuon events and we call this ratio R. The calculation is straightforward: we just need to add the charge squared for all the different possible quark pairs. In fact the up quarks provide $(2/3)^2 = 4/9$, down quarks $(-1/3)^2 = 1/9$ and strange quarks $(-1/3)^2 = 1/9$ which sums to 6/9, and as the muon charge is -1 this squared is just 1. The ratio R between hadronic events and muon pairs should be 6/9 if this calculation is correct. When the experimental measurements were made, the observed ratio R was found to be much higher than this. However, we have forgotten to allow for the three different colours of each quark. Although these have the same charges we must count them properly into the sum. The muon is a lepton and has no colour charge so the ratio should be increased by a factor of three to give $R = 6/9 \times 3 = 2$, which is in good agreement with the experimental data.

To summarise, we are now convinced from a wide range of experiments that quarks exist. Each quark carries a colour charge which is one of only three different colours. The interquark force is carried by coloured gluons. The only states containing quarks which exist in nature are colourless combinations of three quarks or a quark–antiquark pair.

3.8 Are quarks and leptons related?

We have now summarised the most recent knowledge on the structure of matter. All matter is either in the form of leptons or hadrons which are in turn composed of quarks which cannot exist in isolation. Neither quarks nor leptons appear to have any internal structure, at least down to the distances that can be probed with current accelerators. We cannot predict the experimental results of the future but at present these do seem to be the building blocks of matter. The electron has remained as an elementary particle even

Charge	−1	0	$+\frac{2}{3}$	$-\frac{1}{3}$
	e^-	ν_e	u	d
	μ^-	ν_μ	c	s
	τ^-	ν_τ	t	b

Figure 3.9. A table of the known quarks and leptons arranged in three generations. The charged leptons are arranged in order of increasing mass with a separate type of neutrino for each lepton type. The quarks are also arranged in order of increasing mass. When the three different colours of quarks are included the sum of the electric charge of the quarks and leptons in each generation becomes zero.

though it has been studied in detail for almost a century. Perhaps the quark will prove as difficult to split. We just do not know.

There is some circumstantial evidence that the quarks and leptons are related. We can arrange the known leptons and quarks in columns of common charge and in rows (generations) of increasing mass (figure 3.9). The first generation includes the electron, together with the up (charge $+2/3$) and down (charge $-1/3$) quarks, which are the constituents of all the atoms that exist naturally. The electron-type neutrino,ν_e, which is emitted in β decay, completes the first generation. There are three known 'generations' of charged leptons, the electrons, muon and tau, each with its own associated neutrino. We must also remember that each type of quark can carry one of the three different colours whereas the leptons do not carry colour charge. If we add up the charges of the leptons and quarks in any generation we find a sum of zero, providing we remember to include the three different colour species of each quark. This sum has contributions from the neutrino (0), electron (-1), up quark $(3 \times 2/3)$ and down quark $(3 \times -1/3)$. In many of the theories of particle physics this exact cancellation of charge produces a great simplification. The fact that the quarks and leptons can be arranged to produce this cancellation may just be a coincidence, but it is certainly a remarkable one.

We have already met the up quark and the down quark from the first generation and the strange quark (charge $-1/3$) which is associated with the second generation of figure 3.9. During the 1970s particles containing the 'charm' quark (charge $+2/3$) and the 'bottom' quark (charge $-1/3$) were discovered, each with progressively larger masses. Very recently, in 1984, preliminary evidence for the 'top' quark (charge $+2/3$) was observed. When these quarks are added in order of increasing mass to the known leptons, then a symmetric list of quarks and leptons is obtained. All the known particles can be constructed from these three generations of quarks and leptons together with their antiparticles. Whereas the quark theory required just three colours for quarks, the number of generations of quarks and leptons is not so readily predicted or checked.

Each generation of two quarks and two leptons seems to form a logical grouping in current theories of particle physics. In the next few years there will be some crucial measurements, described in chapter 14, which should put a firm upper limit on the number of massless, or light, neutrinos that exist. If each generation continues to include such a light neutrino then these measurements will provide a measurement of the number of generations. There are already strong cosmological arguments, based on the ratio of hydrogen and helium observed in stars, that there are only three or four generations of quarks and leptons. Direct measurements of the properties of the Z boson should soon provide independent results on this important question.

3.9 Summary

In this chapter we have briefly described some of the important ideas and results that have been obtained from experiments searching for the smallest particles inside the atom. We could easily have spent several chapters introducing these interesting new ideas more comprehensively but this would have left less room for the main part of the story. The aim has been to introduce the main ideas that we will need in order to understand the search in later chapters. There are many good books which describe the discoveries of quarks and leptons in more detail. There is now incontrovertible evidence for quarks inside the proton and neutron from three different types of experiment, although single free quarks have not been observed even in the most energetic collisions. These up and down quarks have charges of $+2/3$ and $-1/3$ and each appears with three different colour charges. It seems that objects which carry colour charge cannot exist alone. High-energy collisions have now produced particles which are composed of heavier strange, charm and bottom quarks and very recently preliminary evidence for the top quark has been found. Each of these quarks carries a charge of $+2/3$ or $-1/3$, just like the up and down quarks.

The electron has been joined by two other similar, but heavier, charged leptons, the muon and the tau. Finally a neutrino of each lepton type completes the known leptons. The leptons and quarks can be arranged into three generations of particles in order of increasing mass. The first generation, which includes the electron, up and down quarks, contains the building blocks of all naturally-occurring atoms. There are no precise predictions of the total number of generations of quarks and leptons, although cosmological evidence favours just three generations and detailed particle physics measurements of this number are expected in the next few years.

Finally, all known particles have antiparticles which have the same mass but opposite charge to the particle. Even neutral particles have antiparticles but in this case the baryon or lepton number is opposite to that of the particle.

Chapter Four

Unification of forces

4.1 Are the fundamental forces related?
A common aim of all science is to explain as many facts as possible with a few simple principles or ideas. This leads to efforts to relate apparently different phenomena wherever possible, and in some cases a unification is achieved. We find in the study of fundamental forces that forces which are at first sight very different are indeed closely related. We shall see that the W and Z bosons were predicted as a result of the idea that the electromagnetic and weak nuclear forces derive from a single force. It should be stressed again that these forces have very different properties. The range of the electromagnetic force is infinite and that of the weak nuclear force only 10^{-17} metres and so the linking of these two forces seems at first sight to be highly unlikely. In order to introduce the idea of unification let us return to the early part of the last century.

At that time the studies of electricity and magnetism seemed to be investigations of two different phenomena. In electricity the experiments involved charges and currents, whereas in magnetism the effects were produced by magnets. The earliest studies of magnetism investigated the properties of bar magnets, whereas those of electricity dealt with static charge, and no relation between these topics had appeared. Later studies with moving charges, or currents, showed that an electric current was generated in a wire when a magnet was moved in its vicinity. It was also found that an electric current always produced an associated magnetic field and so these effects seemed to be interrelated. There were, however, some interesting differences between electricity and magnetism. Although single charges could be isolated, this was not possible for magnetic poles. A simple bar magnet has a north and south pole, where the names arise from the

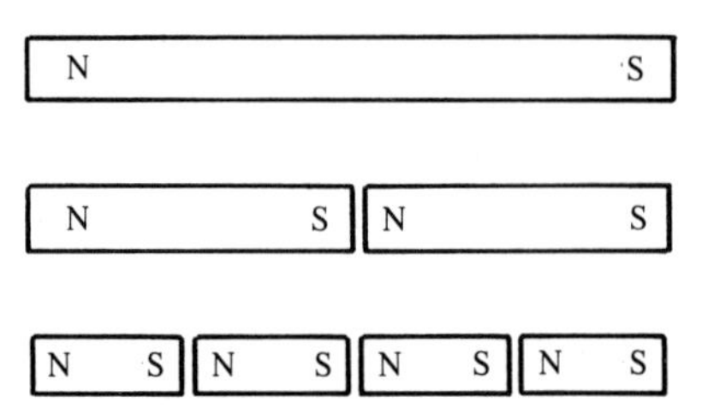

Figure 4.1. Whenever a magnet is broken a new pair of north and south poles is produced and a single magnetic pole has never been observed. This is quite different to electric charge where a single charge can be isolated.

direction in which a magnet points when freely suspended, due to its interaction with the magnetic field of the earth. When a magnet is broken, two further poles are created (figure 4.1) and it is not possible to create a single magnetic pole. The properties of bar magnets are now understood to be due to the electric currents of the electrons in the atoms of the magnet. These effects become measurable because the large numbers of atoms in the magnet interact with each other to produce a correlated effect, whereas in most materials the small effects just cancel out in a random manner.

Nowadays all of these effects are described by one theory of electromagnetism which James Maxwell introduced in 1865. This unifies these very different topics in a single beautiful theory. It is by direct analogy with this successful unification that the electroweak theory manages to embrace both the electromagnetic and weak nuclear forces. However, we should first introduce the topic by discussing Maxwell's unification of electricity and magnetism.

4.2 The electromagnetic interaction

Maxwell collected together the mathematical formulae that had been developed to explain the observations in electricity and magnetism. He noticed that there was an inconsistency in the equations which could most simply be corrected if a changing electric field generated a magnetic field. The reverse process where a changing magnetic field generates an electric field was already well known. This predicted new effect, when subsequently discovered, was a great success for Maxwell's theory of electromagnetism. This is one of the most important theories in physics. The four equations of Maxwell's theory can be used to describe all aspects of electricity and magnetism involving static and moving charges and magnets. However, this economical theory had even more surprises in store.

Once it had been established that changing magnetic fields could produce an electric field and that changing electric fields could produce a magnetic field a remarkable new possibility arose. These changing fields were a self-sustaining method of carrying energy through space. Electromagnetic waves consist of changing electric and magnetic fields which are able to transmit energy, even through a vacuum, and over vast distances. The solutions of Maxwell's equations were of the form of transverse waves where changes occur perpendicular to the direction in which the wave is travelling. These waves had a velocity which could be calculated from measurements of other effects in electricity and magnetism. The calculated velocity for these waves was 186000 miles per second or 3×10^8 metres per second. So this unification of electricity and magnetism also predicted the existence of electromagnetic waves travelling at the speed of light.

Visible light was a well-studied phenomenon at that time and was known to consist of transverse waves and to have a velocity consistent with that of the predicted electromagnetic waves. Perhaps light was an electromagnetic wave? But the solutions of Maxwell's equations predicted waves with all wavelengths. Remember that the wavelength is the distance between two successive crests in a wave, which for visible light is in the range of a few hundred nanometres (10^{-7} metres). So another prediction follows from this unification of electricity and magnetism. Other electromagnetic waves should exist, also travelling with the same speed as light. Later quantum theory developments showed that these waves could be considered as many packets of energy called photons. So the search for electromagnetic waves was also the search for photons. In this story we will be studying in some detail the recent search for the W and Z bosons, which were predicted from the unification of the weak nuclear and electromagnetic forces.

It was 20 years later before the German Heinrich Hertz in 1888 confirmed Maxwell's prediction in every detail. Two separated coils were used to act as transmitter and receiver. An induction coil produced a pulse of high voltage which charged up a wire loop until the voltage exceeded the breakdown voltage in air. Sparks were then produced in the transmitter coil and these generated electromagnetic waves. When the changing magnetic field of these waves reached the coil of the receiver it generated an induced voltage which created sparks in the receiver. The changing magnetic field was shown to be transverse to the direction of the wave by rotating the plane of the second coil so that it was perpendicular to its original position, when no sparks were produced in the receiver. Hertz performed a number of experiments to demonstrate that the effects were due to transmitted waves and measured their wavelength to be around 6 metres. This detection of the predicted electromagnetic waves closed an important chapter in the understanding of physics and was a triumph for the unified theory of electromagnetism.

Although this work was completed in the last century, there have been more recent developments in this area, due to the impact of quantum mechanics and relativity. When applying the electromagnetic theory to atoms and smaller particles the effects of quantum mechanics become very important. The uncertainty principle showed that virtual photons can be emitted for a short time and these can be considered as carriers of the electromagnetic force. Once particles are travelling with speeds approaching that of light, relativity predicts that they behave in a very different way to those objects that we normally encounter, which are travelling at much lower speeds. The mass of a particle gets larger as its speed increases and even its lifetime, as measured by an observer at rest, is increased if it is moving quickly.

Any exact theory of electromagnetism has to include these effects and by the 1940s a complete theory had been constructed by Richard Feynman, Sin-itiro Tomonaga and Julian Schwinger. This theory is called quantum electrodynamics (QED) and has been phenomenally successful in making detailed predictions of electromagnetic effects. The ideas we have been developing of the atomic world with virtual photons and the uncertainty principle are not just approximate ways of explaining the behaviour of atoms. When taken with very detailed mathematics these ideas can be used to predict measurable electron properties with very high accuracy. The agreement between theory and experiment is better than one part in a million million; in fact QED is the most precise theory in the field of physics. This gives us great confidence that the basic ideas are very sound, and leads us to consider whether similar ideas can be applied to the understanding of nuclear forces.

4.3 Weak nuclear force

Before we can usefully examine the ideas involved in linking the electromagnetic and weak forces we must first summarise the information that has been accumulated on the weak nuclear force over the past 40 years. This will necessarily be brief and highly selective but we will try to extract the features that are relevant to the development of the electroweak unified theory.

In the radioactive decay of a nucleus which produces β particles, the weak nuclear force is responsible for the breakup of a neutron. The neutrons, which are constituents of the atom in all elements except hydrogen, are usually stable. However, in some radioactive elements it is energetically favourable for the neutron to decay into a proton, electron and an antineutrino. When this decay occurs, the proton remains in the nucleus and the electron (or β particle) is emitted from the nucleus, together with the unseen non-interacting antineutrino. This same weak nuclear force was later found to be responsible for several other decays, including the decay of a muon to an electron, an antineutrino and a neutrino:

$$\mu^- \to e^- + \bar{\nu}_e + \nu_\mu.$$

Further experiments showed that the weak nuclear force has a universal interaction strength, like the electromagnetic interaction, and also a very short range of just 10^{-17} metres. The most natural way to explain this short range is to assume that the carrier of the force is very heavy. Assume that the intrinsic strength of the weak nuclear force is the same as the electromagnetic force but that its short range reduces its effective strength. Then the mass of this force carrier, the W boson, can be estimated to be 100 times the mass of a proton. In order to explain the known decays and reactions mediated by the weak nuclear force, the W boson has to exist in both positively and negatively charged states. In 1957, the β particles in the decay of cobalt nuclei were found to be preferentially emitted in the direction of the nuclear spin. This was the first observed example where a decay or a reaction was different from its 'mirror' reaction. A 'mirror' reaction is a reaction involving the same particles as would be observed in a mirror. When these reactions are not identical then we say that parity is violated. Later checks showed that parity violation only occurred in the weak nuclear interaction and for this force, the violation was always as large as possible! This has some very interesting consequences for our story.

The massless neutrinos are uncharged and as they are leptons only participate in the weak nuclear interaction. This means that all neutrinos are created by the weak nuclear interaction and so they are most affected by parity violation. Most particles have their spin pointing equally in all directions. However, all neutrinos have their spin pointing against their direction of motion and we call them left handed. On the other hand, antineutrinos always have their spins pointing along their direction of motion and are called right handed. As the particles are massless they also travel at the velocity of light, independent of their momentum, and so these particles are certainly very unusual.

We have seen that the electromagnetic force is carried by a virtual photon, which is uncharged. However, all the early interactions of the weak nuclear force involved the exchange of a charged W boson. These interactions, where charge is exchanged, are known as charged current weak interactions. As there were no example of neutral current weak interactions, where no charge is exchanged, it seemed very unlikely that the electromagnetic and weak nuclear forces could be unified into a single theory.

4.4 The search for weak neutral currents

During the 1950s and 1960s extensive efforts were made to identify the first example of a neutral current weak interaction. These experiments were conducted almost exclusively by searching for the decay of the neutral kaon to two muons. This seemed to be an ideal way to identify this neutral current interaction because this decay (figure 4.2) could not be mediated by a photon as the kaon carries strangeness which cannot be carried by the photon. We will call the particle that mediates this neutral current weak interaction the Z boson, which obviously must be uncharged. No such decays were observed, and as the experiments became even more sophisticated, this decay could be excluded to a rate of less than one in every 100 million other kaon decays. It certainly looked as if there were no neutral current weak interactions.

In the early 1970s, energetic neutrino beams became available and these were a powerful new way to investigate the weak nuclear interaction. These ν_μ beams were generated from the decays of muons which were produced in high-energy proton collisions. All known interactions of this neutrino converted it into a muon and this charged current weak interaction was carried by a virtual charged W boson (figure 4.3). Many of these reactions had been identified, but in 1973 at CERN the first evidence for a weak neutral current

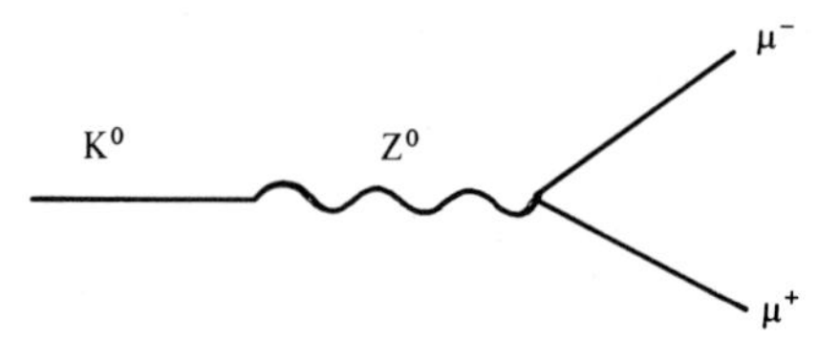

Figure 4.2. A schematic view of the expected decay of the neutral kaon into a muon–antimuon pair. Many experiments searched for this process in the hope of finding the first evidence for a weak nuclear force carried by a neutral particle (Z^0 boson). However, this decay does not occur, even though the Z^0 boson exists, because of an exact cancellation in this process. The first evidence for a neutral current weak interaction was later found in neutrino interactions.

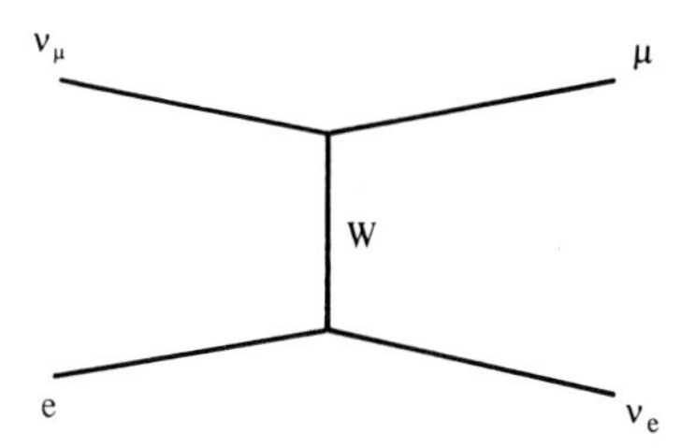

Figure 4.3. A charged current weak nuclear interaction where an incoming neutrino (ν_μ) interacts with an electron and produces a charged muon and an electron type neutrino (ν_e). This charged current interaction is carried by a charged W boson.

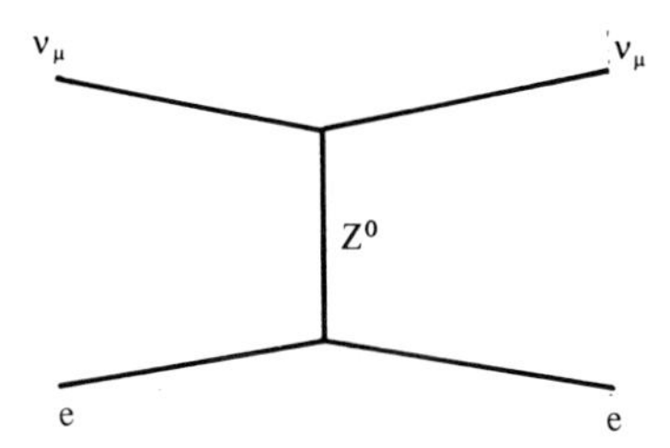

Figure 4.4. A neutral current weak nuclear interaction where an incoming neutrino (ν_μ) interacts with an electron and both particles preserve their identity The neutral current interaction is transmitted by a neutral Z^0 boson. Neither the incident nor outgoing neutrino are visible in a detector and so it was a major achievement when this type of interaction was finally identified.

interaction was discovered. In bubble chamber photographs, interactions were discovered for the first time which did not produce a muon. After extensive checking the conclusion was unavoidable. These neutrino interactions occurred with the neutrino preserved, as expected for a neutral current weak interaction (figure 4.4).

Subsequently it was found that these long-awaited neutral current events occurred in neutrino interactions at a rate of one-third of that of the charged current reactions that had been studied carefully for many years. It is clearly very difficult to identify an event with an unobservable incoming and outgoing neutrino. However, it is a timely reminder that we should always be on the lookout for the unexpected in any experiment. Now the way forward to the unification of the electromagnetic and weak nuclear forces was clear. One question remained, how could we understand the non-observation of neutral currents in the years of careful experiments on kaon decay? The answer to this question led on to the 1974 'November revolution' and the discovery of the charm quark.

4.5 The discovery of charm

With just the three up, down and strange quarks (u, d, s) it seemed impossible to explain the presence of neutral current interactions in neutrino interactions, and their absence in kaon decays. In 1970 S. Glashow, J. Ilioupolos and L. Maiani suggested an elegant solution to the problem, which involved the introduction of a fourth, unobserved charm quark (c). With this extra quark the kaon decay to muons could be made to disappear by an almost exact cancellation. This cancellation only occurred in processes involving a strange particle like the kaon, and to achieve the vanishing decay rate to muons the charm quark could not be very much larger than that of the strange quark. In fact it was forced to be less than twice the mass of the proton.

The weak nuclear force interacts with both quarks and leptons. In charged current weak interactions a muon-type neutrino (ν_μ) always converts to a muon and in a similar way the other types of neutrino, ν_e and ν_τ always convert to an electron and tau lepton respectively. To emphasise this the leptons are frequently arranged in separated pairs or doublets as shown in figure 4.5. The charged current weak nuclear interaction, mediated by the charged W boson, simply changes one particle in a doublet to the other. When we consider the charged current weak interaction for quarks we find a similar picture. Most of these weak interactions with an up quark (u) convert this quark into a down quark (d) or vice versa. The charged W boson changes the quark identity between this pair of quarks. The strange (s) and charm (c) quarks form a similar doublet and the charged W boson changes one of these quarks to the other (figure 4.5).

In November 1974, in two laboratories on opposite coasts of the USA, an unusual new particle later called the J/ψ was found. The

Leptons   Quarks

$$\begin{pmatrix}\nu_e\\ e\end{pmatrix}\begin{pmatrix}\nu_\mu\\ \mu\end{pmatrix}\begin{pmatrix}\nu_\tau\\ \tau\end{pmatrix}\qquad\begin{pmatrix}u\\ d'\end{pmatrix}\begin{pmatrix}c\\ s'\end{pmatrix}\begin{pmatrix}t\\ b'\end{pmatrix}$$

Figure 4.5. In the charged current weak interactions, carried by the W boson, the leptons always convert from a charged lepton to a similar type of neutrino or vice versa. There is no evidence for any mixing between the different lepton types and to emphasise this the leptons are grouped into three pairs. A similar but not exact result is obtained for the quarks, which also convert primarily within a single doublet, but the lower quarks are shown as d′, s′ and b′ as a reminder that this is an approximation.

unusual name is a reminder that the particle was discovered simultaneously in two laboratories and called J and ψ on the east and west coasts respectively. This was the first particle ever detected which contained a charm quark. In fact it contained both a charm and anticharm quark, each with a mass around 1.5 times that of a proton. Although many unstable particles have been discovered over the years, the J/ψ caused special excitement. This was partly because it had been predicted rather accurately, completing one part of the unification story, but also because it had very unusual properties. This particle had a mass of 3100 MeV, three times heavier than a proton, and yet it had a very much longer lifetime than other particles of such high mass. How could it exist for so long before it decayed? The most natural way for a particle containing a charm quark and antiquark to decay is into two particles, one taking the charm quark and the other the antiquark. Later studies showed that the lightest particle containing a single charm quark has a mass of more than 1850 MeV so this simple decay to two such particles is clearly excluded, because they are too heavy. This is the origin of the relatively long lifetime of the J/ψ.

So the prediction of the charm quark was verified by experimental measurements in 1974 and the weak neutral current interactions could at last be understood. The new charm quark prevented the two-muon decay of the neutral kaon and allowed the observed neutrino neutral current interactions. With hindsight it seems incredible that the kaon decay received all the early attention but there were no neutrino beams in those early days.

4.6 What is gauge invariance?

One of the important ideas we meet in physics is that of the conservation law. Some of these laws seem to be exact so what are their origins? A conservation law arises from a symmetry of nature and there are many familiar examples. If we insist that the result of an experiment is independent of the place that the experiment is done, then we can show that exact momentum conservation must be obeyed. In the same way, if we insist that physics measurements at different times give the same results, then exact conservation of energy is an unavoidable consequence. All successful theories of fundamental forces are gauge theories but what does this mean? We have seen in the quantum world that the exact position of an electron cannot be defined. However, the probability for finding an electron can be calculated and depends on the 'wave function' of the electron. This wave function provides the most complete description of the

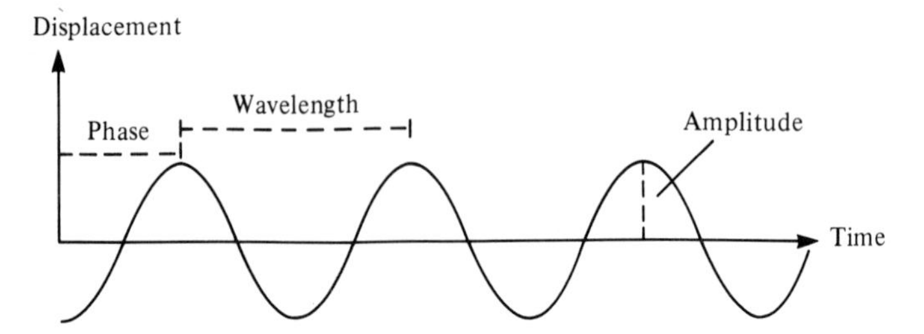

Figure 4.6. The displacement versus time curve for a typical sinusoidal wave illustrates the various parameters of any wave. The amplitude measures the size of displacement at a maximum, the wavelength is the distance between successive peaks and the phase is related to the distance between the first peak and the displacement axis. For all waves the velocity is equal to the frequency times the wavelength.

particle and this can be represented as a wave. We have already discussed wave–particle duality, and so this should not be too much of a surprise. A uniform wave is completely defined by its amplitude, wavelength and phase (figure 4.6). The phase determines the position of the wave and changing the phase is equivalent to moving the wave to the left or right. Although this discussion is getting rather abstract, keep thinking because we have an important point to make very soon. To recap, the key points are that any particle can be represented by a wave and the position of each wave depends on its phase. As discussed earlier, physics results will be independent of the time of the measurement if energy is conserved. Imagine that we required any physics result to be independent of making a single phase change to the wave functions of all particles. We find that the necessary consequence of this requirement is that exact charge conservation must be obeyed. This means that the sum of the charges before and after any interaction must always be the same and this is what we find to be true in nature. So perhaps these phases are more useful than you thought. Suppose we are more ambitious and ask that all physics measurements should be independent of an even more general change in the phase of our particle waves. We will allow different phase changes at all places and times in the universe. In other words we can choose different origins for our waves at each space point and time in the universe. This is called local gauge invariance and despite its name is the most powerful invariance principle of all. Just imagine how complicated it is to keep every experimental measurement unchanged, while still allowing completely arbitrary changes to the phase of any particle waves at all points of the universe. These experiments can include interference experiments, which are very sensitive to the relative phase of the waves, so how can this result be obtained? What is the consequence of requiring this local gauge invariance for the electromagnetic interaction? The invariance can only be achieved by introducing an electromagnetic field with the following properties. This field must be carried at the speed of light and interact with charges in a universal and predictable way. Remarkably we find that this is the exact form of interaction which we measure in electromagnetic experiments. The existence of the photon and its properties can all be predicted by requiring local gauge invariance. In effect, we can also explain Maxwell's equations by using this same idea. To summarise, the theory of quantum electrodynamics is a gauge theory which satisfies local gauge invariance and is able to predict experimental results to phenomenal accuracy.

Consider the electroweak theory, which is the unified theory of weak and electromagnetic interactions. What does the application of this same local gauge invariance principle predict there? The most important technical difference when applying these ideas to the electroweak theory is that the force carriers now have mass. The force carrier of the electromagnetic force is the photon, which is massless, and this leads to a great simplification. As we are interested in applying local gauge invariance to the massive W and Z bosons, some new ideas are needed. Somehow the mass has to be cleverly included in the theory, without losing all of the symmetry, and this process is called spontaneous symmetry breaking.

Its details are beyond the scope of this book, but its consequences are not. This mechanism requires the addition of four massive new objects into the electroweak theory. Three of these provide the mass for the W^+, W^- and Z^0 bosons, without destroying the symmetry, but there remains one observable particle called the Higgs boson. This particle is named after Peter Higgs of Edinburgh University who introduced this important idea; it is predicted to be uncharged and to have zero spin. Its interactions with all other particles can be calculated, but unfortunately its mass is not predicted.

There was one last crucial theoretical step to be solved before the electroweak theory could be completed. In the early days of quantum electrodynamics, detailed predictions had been impossible because infinities kept occurring in the mathematical calculations. This was later solved by the idea called renormalisation, which successfully avoided these problems. However, the more complicated electroweak theory, where the force-carrying W and Z bosons can interact with each other, was not proven to be a renormalisable or calculable theory. This crucial result was finally proved for the electroweak theory by Gerard t'Hooft. At last the powerful ideas of local gauge invariance could be applied to the electroweak theory.

So, to summarise, a local gauge theory has been phenomenally successful in predicting the most detailed results involving the electromagnetic force. This same idea could be applied to the electroweak theory, once the problems of the force carriers having mass and renormalisation had been solved. The massive force carriers of the weak nuclear force have one other important difference from the photon. They carry a new type of charge, let us call it a weak charge, and so are able to interact with each other, whereas the photon is uncharged and can only interact with electric charges and not another photon. Gauge theories which involve a force carrier like the photon which does not itself carry electric charge are easier to calculate theoretically. When the force carrier can interact with another force carrier this leads to some interesting new predictions in the theory.

4.7 What is needed for the electroweak theory?

We saw for the electromagnetic force that local gauge invariance required a photon to exist with well-determined properties and interaction with charges. In the electroweak model, local gauge invariance determines the detailed form of the interaction of the carriers of the electroweak force. Sheldon Glashow, Abdus Salam and Steven Weinberg were awarded the Nobel prize for physics in 1979 for their contributions to the standard model of weak and electromagnetic interactions or electroweak theory. This must contain charged W^+ and W^- bosons and at least two neutral force carriers to provide the electromagnetic interaction and the neutral current weak interaction. If one of these four carriers was the photon, then the electromagnetic and weak nuclear forces would not be intimately related.

We have seen that each force interacts with a certain property of matter. So far we have considered the gravitational force interacting with mass, the electric force with electric charge and the interquark force with colour charge. Up until now we have described the weak nuclear force as interacting with a weak charge. However, on closer inspection we find that there are two different types of weak charges, known as weak isospin and weak hypercharge. The weak nuclear force is carried by two different force carriers which each interact with just one of these types of weak charge. The W bosons carry the force between the weak isospin of particles and a new particle called the B boson carries the force between weak hypercharges. In this theory the W bosons can carry electric charge and are expected to occur as W^+, W^- and W^0 states, whereas the B^0 boson carries no electric charge. Although these new properties have unfamiliar names, they are just another property of a particle, like its electric charge. These properties are introduced to explain the behaviour of particles in the presence of other particles or electric and magnetic fields.

The strength of the interaction between a particle and the B boson depends on its weak hypercharge. No particle has a weak hypercharge of zero and so the B boson interacts with every right- and left-handed particle or antiparticle. We have seen earlier that all neutrinos are left handed, and have their spins pointing against their motion, but both right- and left-handed leptons and quarks exist. In experiments at very high energy, the mass of the quark or lepton becomes negligible compared to their kinetic energy, and then the weak nuclear force exhibits a very unusual property. The charged W bosons only interact with left-handed quarks or leptons. The right-handed quarks and leptons do not 'feel' any force carried by the W bosons. The weak isospin of left-handed quarks and leptons is $+1/2$ or $-1/2$, but the weak isospin of right-handed quarks and leptons is zero to be consistent with this unusual property of the W bosons.

You will have noticed that in this wonderful new theory we have lost the photon and the Z^0 boson! The final dramatic result is obtained by realising that the W^0 and B^0 are not the states which appear in nature. These states, which each carry part of the electroweak interaction, mix together to produce the observable photon and the Z^0 boson. This is illustrated in figure 4.7 where the angle, θ_w, is called the Weinberg angle. The photon and Z^0 boson which appear

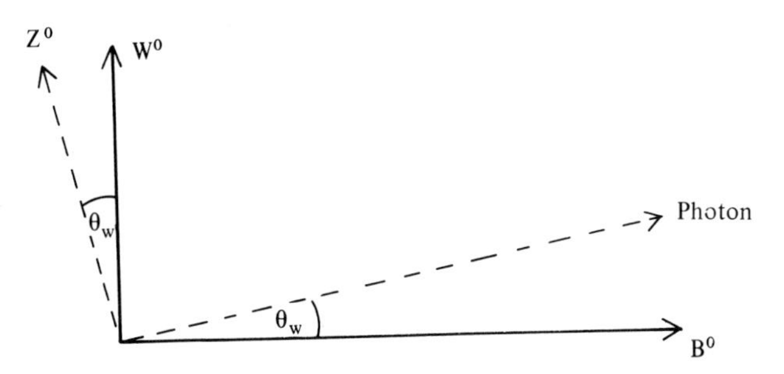

Figure 4.7. This emphasises the intimate relationship between the photon and the Z^0 boson. The neutral carriers of weak isospin (W^0) and weak hypercharge (B^0) do not appear as observable particles. They mix together to produce the photon and Z^0 boson and the mixing is governed by the angle θ_w known as the Weinberg angle. The relations between these four objects can be derived simply from this figure.

in nature are an intimate mixture of the electromagnetic and weak neutral current interactions. Local gauge invariance predicts the detailed properties of the W^-, W^+, W^0 and B^0 bosons. However, the exact form of the mixing between the B^0 and W^0, which produces the observable photon and Z^0, is not predicted. This undefined parameter of the electroweak theory is usually written as $\sin^2\theta_w$.

As this parameter enters most calculations of processes involving the electroweak interaction it can be determined from many different types of experiment. Over many years the value of $\sin^2\theta_w$ has gradually been determined more accurately and it is currently measured as $\sin^2\theta_w = 0.23 \pm 0.01$ from an average of many experiments.

Now the detailed properties of the W^-, W^+ and Z^0 bosons can be predicted from the unified theory of electromagnetic and weak interactions. The masses of the W^+, W^- and Z^0 bosons can now be predicted in terms of measured quantity $\sin^2\theta_w$. The naive predictions for the electroweak model yield mass values for the W^+ and W^- bosons of $(37.3/\sin\theta_w)$ GeV which is approximately 78 GeV. The corresponding prediction for the mass of the Z^0 boson is $M_w/\cos\theta_w$ which equals 89 GeV. As the mass of a proton is around 1 GeV both of these bosons are nearly 100 times as massive as the proton and so it requires enormous energy to create them as real particles. As the weak nuclear force is active in the sun and radioactivity we know that virtual W and Z bosons are at work all around us. But without the provision of enough energy to provide their real mass, they could never be directly observed in such reactions.

When the predictions for the masses are calculated in the most precise way both the W and Z masses are increased by a few per cent. This increase is mainly due to the self-interaction of the bosons that we discussed earlier (p. 54). This happens because these carriers of the weak force also carry a weak charge and so can interact with each other. The details of this interaction are precisely calculable in the theory and so an accurate measurement of the masses of the W^+, W^- and Z^0 bosons would check very detailed elements of the gauge theory of the electroweak force. The electroweak theory is a beautiful theory which involves many difficult mathematical steps, but most of the main physical principles have been mentioned in this brief discussion.

4.8 Summary

In this chapter, which contains by far the most difficult concepts in the book, we have had to condense an enormous number of experimental results and theoretical ideas down to a few crucial facts. Some of the ideas are very difficult and have taken many years to develop, so do not be discouraged if some of them require some additional reading. There are many other ideas or experiments that could have been added to this book to make the treatment more comprehensive. However, the aim has been to present the key concepts and results that led to the predictions of the existence of massive W and Z bosons, without including a full primer on particle physics, quantum mechanics, relativity and gauge theories.

The electromagnetic interactions were successfully unified by Maxwell and his predictions of electromagnetic waves (photons) travelling at the speed of light were verified by Hertz. Many detailed experiments on the weak nuclear force showed that it had a common strength in different interactions, a very short range and also exhibited parity violation, unlike other forces. For this reason only left-handed neutrinos exist in nature. All observed weak nuclear interactions had involved a charged current between the interacting particles, until the weak neutral current interaction was discovered in neutrino interactions in 1973. The mystery of why this interaction did not show up in the decay of the neutral kaon was solved by the introduction of the charm quark. The J/ψ, which is made up of a charm quark–antiquark pair, was discovered in 1974.

The principle of local gauge invariance was an important theoretical development. This principle seems to control the form of the various interactions and allows detailed predictions to be made. However, as the W and Z bosons are massive, the idea of spontaneous symmetry breaking had to be introduced before the gauge principle could be successfully applied. The most popular method for doing this brings in a new massive Higgs particle, but unfortunately its mass is not predicted. The final steps for the electroweak model included the proof that the theory was renormalisable and the choice of groupings for the particles and the carriers of the weak force. The simplest choice, which was consistent with the known properties of the weak interaction, is known as the standard model of weak and electromagnetic interactions or electroweak theory. This predicts the existence of the W^+, W^- and Z^0 bosons with masses around 100 times that of a proton. These predictions are very precise and include detailed properties of expected masses and decay properties. This theory emphasises the intimate link between the photon and Z^0 boson which result from a mixing of two unobservable force carriers. The rest of this story is about the experimental search to check these predictions.

Part Two

The objective

Chapter Five

The experimental test

5.1 The objective

We have seen that the electroweak theory provides a link between the electromagnetic and weak nuclear forces. This is a remarkable achievement as it explains naturally many experimental observations. If correct, it means that the electroweak force, together with the gravitational and strong nuclear forces, are the only known fundamental forces of nature. This unification of forces appears to be very sucessful, but it also makes very firm and precise predictions. The electroweak theory predicts that W and Z bosons exist with masses around 100 times that of a proton. It predicts these masses and other properties of the bosons very precisely. As this theory plays a central role in our understanding of forces and our search for further unification, it is obviously important to verify these predictions by experimental measurements as soon as possible.

The W and Z bosons are the two targets of the search described in this story. Science is an experimental subject and even the most appealing and well-tested theories sometimes have to be rejected. New experimental observations often reveal inconsistencies or serious errors in many theories. Will the electroweak theory that we have discussed stand up to the crucial experimental test? Do these W and Z bosons really exist? In the early 1970s while the electroweak theory was being finalised, it seemed as if the important experimental tests would have to wait until the 1990s.

5.2 Production of W and Z bosons

The search for W and Z bosons requires very high-energy accelerators and the energy to produce their very large masses has to be provided by the kinetic energy of accelerated beam particles. The accelerated beam is frequently aimed at a stationary target but in this type of collision there is a net forward momentum which must be preserved after the collision. This means that some of the collision energy has to be used to provide this momentum and is not available for creating new particles. The highest collision energies are achieved by head-on collisions of two accelerated beams. In this case there is no net momentum to be provided and the full collision energy is available for particle production. The search for the very heavy W and Z bosons would certainly have to be made at a colliding beam accelerator.

The first accelerator with enough energy to create a Z^0 boson was expected to be a high-energy electron–positron collider. The largest version of this type of accelerator is currently under construction at the CERN laboratory near Geneva. This large electron–positron collider (LEP) should be completed and ready for its first experiments in 1989. This accelerator will include a ring of 4000 magnets arranged in a circle of 30 kilometres, at depths of up to several hundred metres below the surface. Beams of electrons and positrons will be produced and introduced in opposite directions into the collider, where they will be accelerated to a maximum energy of 50 GeV per beam. These counter-rotating beams of particles will then be maintained at this high energy and will collide at four intersections, where large experimental detectors will record the electron–positron collisions. When an electron collides head-on with a positron, the two particles annihilate, making all their energy available for particle creation. The beam energies can be adjusted so that the electron and positron each carry exactly half of the energy required to create a Z^0 boson (figure 5.1). At this energy each electron–positron annihilation will produce a single Z^0 boson and every particle that is detected will be from its decay. In this ideal environment, it will be possible to measure the detailed properties of the Z^0 boson.

There are several existing proton accelerators which already have energies far exceeding those planned for the LEP in 1989. Why is it

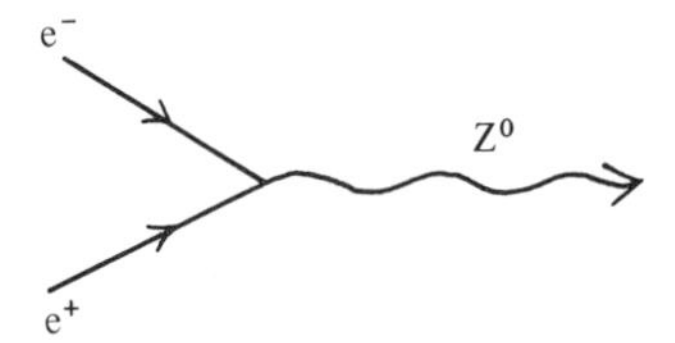

Figure 5.1. An electron–positron collision at just the right energy to create a Z^0 boson. In this case no other particles are produced and so it is an ideal way of studying the various decays of the Z^0 boson.

so much more difficult to accelerate electrons to these very high energies? Whenever a charged particle is deflected it emits a photon to conserve momentum and the number of photons emitted is much larger for the light electron than for the heavier proton. In a circular accelerator, a particle is continually deflected by magnetic fields so an electron is continually emitting photons and losing energy. Very large amounts of electrical power are required in order to maintain the electron beam at high energy. The loss of energy is much less important for the heavier proton and for this reason higher-energy proton accelerators can be operated with much lower power consumption.

Even with the LEP it will be hard to achieve collision energies above 100 GeV with conventional accelerating electric fields. This means that Z^0 production should be possible but what about the W bosons? The W^+ and W^- bosons carry electric charge and so charge conservation prevents the production of a single W boson in an electron–positron collision but they can be produced in pairs as shown in figure 5.2. As the W bosons have a similar mass to the Z boson, it requires twice the beam energy before W bosons can be created in a machine like the LEP!

The plan for the LEP is to have an initial phase starting at the end of 1988 with a collision energy of 100 GeV which is enough to produce the Z^0. Subsequently, development of superconducting accelerating elements, which can produce higher electric fields without more electrical power, will be needed before the energy can be increased. If these developments are successful, observation of the reactions $e^+e^- \to W^+W^-$ should be possible in the mid-1990s. So it appeared that although the theoretical ideas were indeed very impressive, the crucial experimental measurements could only be made after a very long delay.

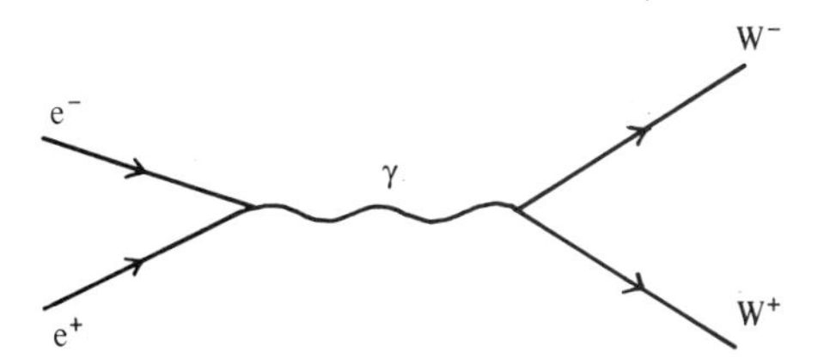

Figure 5.2. The W bosons are charged and so they need to be produced in pairs in electron–positron collisions. The W and Z bosons have a similar mass and so it will require approximately twice as large a collision energy as that needed for Z bosons to produce W bosons at an electron–positron collider.

5.3 A new idea

In 1976 an alternative method of producing the elusive W and Z bosons was proposed by David Cline, Peter McIntyre and Carlo Rubbia. This did not require a new accelerator and could, if successful, yield results within 5 years. This brilliant idea was conceptually very simple and the proposal involved converting a high-energy proton synchrotron to accelerate simultaneously a beam of antiprotons and a beam of protons in the accelerator ring. The antiprotons could be accelerated in the opposite direction and collided with the protons in head-on collisions. As the largest available synchroton had maximum energies of several hundred GeV this would already provide a collision energy large enough to check if the W and Z bosons really did exist.

However, there are several disadvantages compared to using electron–positron collisions. We have seen that each proton is made up of three valence quarks, gluons and quark–antiquark pairs and that each antiproton contains three antiquarks together with gluons and quark–antiquark pairs. When a proton collides with an antiproton it is a very complicated interaction between many different particles. The beam energy of several hundred GeV is very large, but each quark only carries a small fraction of this energy. If a W or Z boson is produced, it will be accompanied by a large number of other particles which are the remnants of the complex collision. The most direct way of producing a Z^0 boson is by the annihilation of a quark from the proton with an antiquark from the antiproton (figure 5.3). The other quarks, antiquarks and gluons need to regroup after the collision into pions, protons and other hadrons.

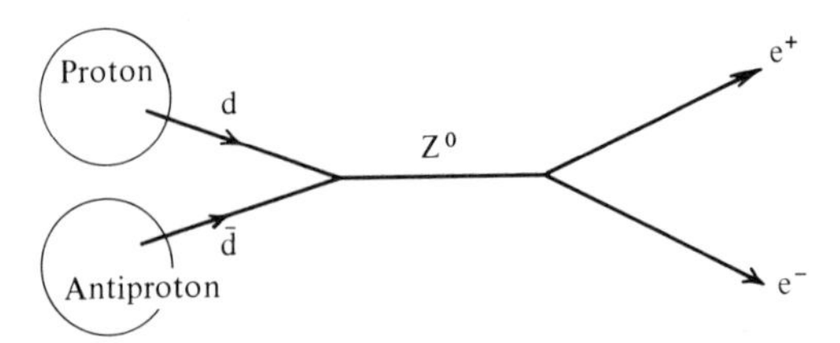

Figure 5.3. A schematic view of Z^0 boson production in a high-energy proton–antiproton collision. Only the most energetic interaction is shown and there will be many additional particles produced in the collision. The annihilation of a quark–antiquark pair $(d,\bar{d})$ produces the Z^0 boson which rapidly decays to lighter particles. In about 3% of the decays an electron–positron pair is produced which is the simplest decay to detect. Alternatively any quark–antiquark or lepton–antilepton pair can be produced providing the particles are not too heavy.

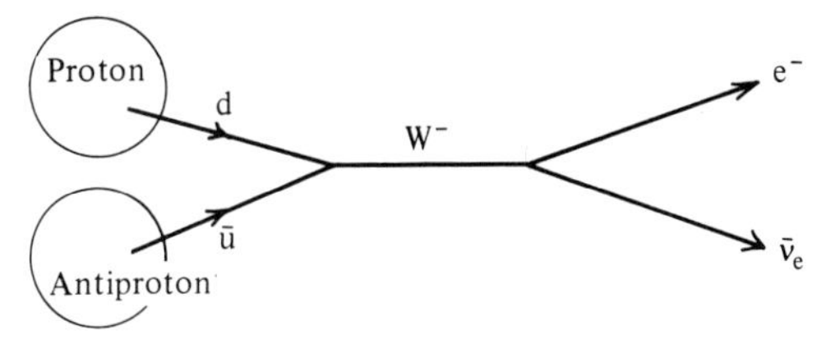

Figure 5.4. There are many constituents involved when a proton collides with an antiproton. In this interaction a d quark from the proton has interacted with a ū antiquark from the antiproton, producing a W^- boson. In a short time this decays to an electron and an antineutrino ($\bar{\nu}_e$), or into two other leptons or quarks. All the other constituents can also interact and so many other particles are produced in a high-energy hadronic collision but this diagram only indicates the most energetic part of the interaction. Notice that a single W boson can be produced in such a collision.

However, there is a major advantage when it comes to the production of charged W bosons. The quarks have charges of $+2/3$ and $-1/3$ and antiquarks of $-2/3$ and $+1/3$. Consequently quark–antiquark collisions can produce objects which have charges of $+1$, 0 or -1 in a single collision. This means that W^+ and W^- bosons can be produced singly like Z^0 bosons in such a collision (figure 5.4). An even more fundamental advantage of the antiproton–proton collider idea was that proton accelerators with energies of several hundred GeV already existed!

With this new idea, an existing synchrotron could be adapted with relatively minor modifications to provide collisions between protons and antiprotons. These collisions would be at energies 10 times larger than had ever been studied at any accelerator. This was a golden opportunity to study collisions at a high enough energy to produce the W and Z bosons and also to search for other new phenomena. However, there were many difficulties and uncertainties to be overcome before the project could become a reality.

There were only two accelerators where this idea could be converted rapidly into a collider project, the CERN laboratory in Switzerland and the Fermi National Accelerator Laboratory (FNAL) in the USA. The new idea could only work if a highly compact beam of antiprotons with a precisely defined momentum could be produced. This was the key element in the whole project.

The first method for compressing proton beams had been invented by Gershon Budker at Novosibirsk in the USSR in 1966 and tests of this idea, known as electron cooling, had been successfully carried out in 1976. In this approach, a beam of electrons was passed in the same direction as a proton beam. When the velocity of the two particle beams was made similar, the spread in the momentum of the proton beam was reduced (cooled). The close contact with the electron beam allows some of the momentum variation from the protons to be shared with the electrons.

The FNAL laboratory began a detailed study of how to utilise electron cooling for the conversion of its accelerator complex to an antiproton–proton colliding beam facility. However, this laboratory had already embarked on an ambitious programme to double the energy of its main proton accelerator from 500 GeV to 1000 GeV. This involved a large research and development programme because it required conversion of all the accelerator magnets to superconducting magnets, which consume less power yet produce higher magnetic fields than conventional magnets.

At CERN a super proton synchrotron (SPS) had been completed in 1976 which could accelerate a beam of protons to an energy of 400 GeV. A different method of beam cooling had been invented by Simon van der Meer at CERN in 1968. This idea, known as stochastic cooling, was not published until 1972. This important publication ended with the modest words 'this work was done in 1968. The idea seemed too far fetched at the time to justify publication. However, the fluctuations upon which the system is based were experimentally observed recently. Although it may still be unlikely that useful damping (cooling) could be achieved in practice, it seems useful now to present at least some quantitative estimation of the effect.'

CERN proceeded rapidly to evaluate the new ideas of beam cooling with a view to the construction of an antiproton–proton collider project and these studies were concentrated on the use of stochastic cooling. In this method, which we will describe later, correcting electric fields are applied to a circulating beam of particles. These are chosen to compress the size of the particle beam and to reduce the spread in its momentum.

As antimatter does not exist naturally on earth, the antiprotons need to be produced. Could enough of them be produced in a reasonable time and kept safely ready for use? The antiproton beam needs to be kept intact for many days and this means that any beam losses have to be minimised. The technology has to be developed and checked so that the beam handling can be performed routinely. The valuable antiprotons, which are accumulated over several days, need to be transferred to the accelerator with very high efficiency.

The antiproton was confidently expected to be as stable as the proton, but it had only been experimentally confirmed to have a lifetime of more than a few microseconds. As this proposal required the beams to be stored for several days, the lifetime of the antiproton was a very important missing number. However, the most important uncertainty involved the cooling of the beam. The antiproton beam had to be compressed into a much smaller space than had previously been achieved. Without an intense beam, it would not be possible to produce enough head-on collisions between antiprotons and protons to have a chance of discovering the W and Z bosons.

5.4 The idea becomes a project

The various ideas had to be collected together into a practical project where all of the unknowns were clearly identified. In this way the risks could be clearly understood and the probability of success could also be evaluated. At the same time a collaboration of people who would be interested in constructing a new detector to observe the high-energy collision needed to be formed. This team then had the critical job of convincing first themselves and then their colleagues that the potential success of such a project was important enough to justify the calculated risks involved.

The new idea of an antiproton–proton collider at CERN was enthusiastically developed and promoted by Carlo Rubbia. His dynamic personality was an important element in the critical early phase of the project, when the idea had to be converted into a practical project. This was helped by his deep understanding of topics as diverse as accelerator technology and theoretical particle physics. Also working at CERN was the brilliant accelerator physicist called Simon van der Meer mentioned above. He had been involved in a number of important accelerator innovations and was working on ways of increasing the intensity of particle beams. Breakthroughs certainly were needed in this area to enable the collider project to become a reality. The directors of CERN, Sir John Adams and Leon van Hove, were also strong supporters of this project.

From the very early days Carlo Rubbia was joined by a number of research groups from many different countries, which formed the underground area 1 (UA1) collaboration. The accelerator is underground and so each detector at the collider had to be installed in an underground area. The collaboration was formed to design and construct a detector to identify and record energetic antiproton–proton collisions and search for the W and Z bosons. Two other major collaborations were formed and proposed to search for W and Z production in different ways. It was finally decided to approve only one more experiment and this became the UA2 collaboration. The decision to have two independent experiments in the search for the bosons proved a wise choice. As these are very complex detectors, it is much safer to have two experiments in case of a major problem with one of them. It also introduced an extra element of competition which ensured that the experimental results were extracted as quickly as possible.

Once the experimental collaborations had formed, the main question still remained. Would CERN proceed with the collider project? In the 1960s CERN had built up a high reputation based on valuable experimental results obtained in many different experiments. The reliability and excellence of its engineering was an important element in this well-deserved reputation. However, many of the really exciting particle physics discoveries continued to be made in the laboratories of the USA. In 1973 the discovery of weak neutral currents at CERN in the study of neutrino interactions was a major boost.

The accelerators at CERN are used by thousands of physicists, mainly from Europe, to investigate the structure of matter and the fundamental forces of nature. There are many very important experiments to be done and so there is fierce competition between different experiments for precious time on the CERN accelerators. Any new experiment that is proposed has to pass a strict selection procedure by committees composed of experienced particle physicists. The choice facing CERN was clear. It could concentrate on its existing accelerator programme and try to ensure the maximum efficiency and running time for its current range of experiments. These were primarily experiments where the high-energy particle beam was taken from the accelerator and made to collide with a stationary target. The alternative option was to continue such experiments but to give high priority to the rapid development of a proton–antiproton collider at CERN. This could only be done within existing budgets by diverting large resources of manpower and equipment to such a programme. As such a new programme would involve much testing and construction, there would clearly be major implications for the many users of the CERN accelerators. The technical challenges in this new programme were demanding and the planned time scale was very short. However, if this collider project could be completed then we would know in the early 1980s whether the W and Z bosons really existed. It was decided to go ahead rapidly at CERN with a series of technical tests that were necessary to check the more difficult ideas involved in the project. When these turned out to be successful, it was decided to push ahead at CERN as fast as possible, to get the collider operational.

We will soon be discussing the accelerators and detectors used in the search for W and Z bosons. Before we do this we need to investigate whether these bosons could be recognised in an energetic antiproton–proton collision. How long will they live and how will they decay? How often will they produced? How can the rare production of W or Z bosons be isolated from the much more numerous 'routine' collisions? In the next chapter these important questions are answered, so that the targets of the experimental search are clearly defined.

Chapter Six

How do we recognise a W or Z boson?

6.1 Introduction

The massive W and Z bosons are clear predictions from the standard electroweak theory and we have already discussed their very large masses which make them extremely difficult to produce. Several questions immediately arise. How would we tell if one was created? Can we make any estimate of how often this will happen in an antiproton–proton colliding beam experiment? How will we distinguish genuine W or Z boson events from other collisions? These and many other questions had to be answered before any experiment to look for these particles could be started. We will now try to construct a picture of how the elusive W and Z bosons may appear, if we have enough energy to produce them. In this way we can find out the basic principles of design for a particle physics detector to identify them in an efficient way.

Up until now we have only encountered the W and Z bosons as virtual particles when they act as carriers of the weak nuclear force and owe their brief existence to the uncertainty principle. Now we are trying to produce them in an energetic antiproton–proton collision, where there is enough energy to create them as real particles.

6.2 How fast will the W and Z bosons decay?

We have already seen that a very short-lived particle can only be detected indirectly by its decay products. The W and Z bosons decay by the weak nuclear interaction so we might have expected that the decay occurs rather slowly. Unfortunately their large mass, which makes them difficult to produce, also means that they decay much more quickly than lighter particles. In a W or Z decay there is an enormous energy release, so there are many available energy states for the decay and it is expected to proceed in less than 10^{-24} seconds! This is an even shorter time than the lifetime of the resonances that we were discussing in a previous chapter and they decayed by the strong nuclear interaction. There is no alternative – we will have to employ the same indirect technique that was used so successfully in the identification of short-lived resonances.

This sounds quite promising but there is immediately a further problem. These high-energy collisions frequently produce over 100 different particles. How can we possibly select the decay products of the W and Z bosons? Fortunately clear predictions were available for the decays of these particles and these will now be described.

6.3 Decay products of a Z^0 boson

The Z^0 boson is neutral and is expected to decay into any lepton–antilepton pair. The most characteristic of these is that into an electron and a positron ($Z^0 \rightarrow e^-e^+$). In this case the heavy particle breaks up into these two light particles and converts all of its mass energy into their motion. If the parent boson is at rest, the highly energetic electron and positron must be emitted with equal energy in a back to back configuration, to conserve momentum. These decay energies are so large that this is a striking and characteristic signal of the presence of a Z^0 boson. However, this decay is only expected to account for 3% of all Z^0 decays and so it would be very useful if other decays could also be identified.

An equivalent number of decays are expected into muon and antimuon ($Z^0 \rightarrow \mu^+\mu^-$). Although the muon is slightly heavier than the electron, they both carry a negligible mass compared to the Z^0 boson and consequently the muon and antimuon should also be emitted with enormous energy in a back to back configuration. The muon is not a stable particle but its lifetime of a few microseconds allows it to traverse an experimental detector before it decays. The muon is the only known charged particle that is highly penetrating and this property can be used in its identification.

The decays into electron–positron and muon–antimuon represent only 6% of the total Z^0 decay possibilities. However, these decay modes are the most useful in a search for its existence in antiproton–proton collisions. The beam particles are composed of quarks and antiquarks and so are most of the particles created in antiproton–proton collisions. Electrons and muons do not feel the strong nuclear

Process/charge			Relative rate
ν	$\bar{\nu}$	0	2
e^+	e^-	1	1
μ^-	μ^+	1	1
u	$\bar{u}$	$\frac{2}{3}$	$3\frac{1}{3}$
d	$\bar{d}$	$\frac{1}{3}$	$4\frac{1}{3}$

Figure 6.1. The predicted rates of Z^0 boson decay into lighter particles are very different for the various quarks and leptons. The rates depend on whether the particle has colour and on its electric charge. The largest decay rate into leptons is into the undetectable $\nu\bar{\nu}$ channel! The decays into quark–antiquark, although more numerous, are much harder to detect than those into charged leptons because similar events can be produced by other processes.

force and are produced only rarely in these collisions. This means that it is easier to recognise the decay into electrons or muons than the much more frequent decays into quarks and antiquarks. The full list of expected Z^0 decays, together with their predicted rates, are shown in figure 6.1. Notice that the decay rates depend on the electric charge of the leptons or quarks, which is a reminder that the Z^0 boson is intimately related to the photon, which exaggerates this dependence by only interacting with charged particles! The quark–antiquark decays are dominant because each quark can exist with three different colour charges. If these decays could only be identified then we would have many more chances of identifying the Z^0 boson. So why do we concentrate on the electron and muon decays?

In the decay of a Z^0 boson to a quark and an antiquark, we do not expect to see these bare quarks in our detector. The strong nuclear force interacts between the colour charges of the quark and antiquark and produces an interesting result. Just as in the production of a quark–antiquark pair in electron–positron collisions, the quark and antiquark 'dress' into hadrons before entering the detector. This dressing occurs over distances smaller than the size of a proton and so no detector can hope to observe this process. As the quark and antiquark move rapidly away from each other the force between the colour charges gets stronger. This colour field can be considered as a string which is stretched until it breaks and each time this happens a quark–antiquark pair is created. In this process extra quark–antiquark pairs are produced which group together to form hadrons such as pions and protons. In a detector the only trace of the original quark or antiquark is a collection of hadrons travelling close together, usually described as a jet. Two energetic, back to back jets of hadrons can be very spectacular but unfortunately these can be produced by other processes than Z^0 decay.

To summarise, the decays into electron–positron and muon–antimuon are expected to be the best way to identify the presence of a Z^0 boson. The more prolific quark–antiquark decays are expected to be much harder to separate from other collision processes.

6.4 Decay products of a W boson

The W boson is a charged particle and both W^- and W^+ are expected to exist. We will consider the expected decays of the W^- boson but the corresponding decays of the W^+ boson are easily obtained by interchanging particle with antiparticle. We can calculate the relative rates of different W^- decays much more easily than for the Z^0 boson because the W boson only interacts with the property known as weak isospin. The decays into leptons can be listed simply by requiring conservation of charge and lepton number in the decay. The only possible leptonic decays are those of W^- into electron–antineutrino ($e^-\bar{\nu}_e$), muon–antineutrino ($\mu^-\bar{\nu}_\mu$) or tau–antineutrino ($\tau^-\bar{\nu}_\tau$). These are each expected to occur in one-twelfth of the W^- decays and all involve the non-interacting antineutrino!

As the W boson is massive these pairs of decay particles will be emitted with very high energies in opposite directions. The energetic electron can be clearly identified so let us concentrate on the most important decay mode of the W boson, namely $W^- \to e^-\bar{\nu}_e$. When we considered the Z^0 decay to electron–positron, both of the decay particles were well-identified charge particles. Even in the most promising decay mode of a W^- boson, an undetectable antineutrino is one of the two decay products. This will leave any detector without interacting and so the detection of W^- decay looks to be very difficult indeed! However, the detectors which were to be constructed to detect the W boson almost totally enclosed the collision point. A non-interacting neutral particle like the neutrino will avoid direct detection but if it is energetic then its presence can be deduced by using momentum conservation. We will discuss this in more detail later but it is sometimes possible to deduce the presence of an energetic neutrino even if it escapes direct detection. The muonic decay of the W^- should also be detectable but it was the decay to electron–antineutrino that was used for the original W boson search.

Process	Relative rate
$e^- \ \bar{\nu}_e$	$\frac{1}{12}$
$\mu \ \bar{\nu}_\mu$	$\frac{1}{12}$
$\tau^- \ \nu_\tau$	$\frac{1}{12}$
d $\bar{u}$	$\frac{1}{4}$
s $\bar{c}$	$\frac{1}{4}$
b $\bar{t}$	$\frac{1}{4}$

Figure 6.2. The predicted decay rates for the W boson can be calculated rather easily. After allowing for the three quark colours there are 12 separate decays that are possible, all with an equal likelihood. The pairs of quarks and leptons are exactly as seen earlier because the W boson is the carrier of the charged current weak interaction.

As the quarks carry three different colours, there are three times as many decays into quark–antiquark compared to those into leptons. The W^- boson can decay into either $\bar{u}d$, $\bar{c}s$ or $\bar{t}b$ quark–antiquark pairs and these quark decays will result in two back to back jets of hadrons just as in the decay of a Z^0 boson. The problem with these decay modes is that this configuration is not exclusively reserved for W and Z bosons and so these are not suitable for the discovery of W or Z bosons. The full list of possible decays of the W^- boson is presented in figure 6.2 together with the expected decay rates.

The initial search for the W boson was conducted for the decay W^- to electron–antineutrino even though this decay includes an 'undetectable' antineutrino ($e^-\bar{\nu}_e$). The other decays are difficult to disentangle from processes which occur much more frequently than W production. The corresponding decay for the W^+ boson produces a positron and a neutrino ($e^+\nu_e$).

6.5 Striking features of W and Z decays

When an antiproton collides with a proton, the constituent quarks and antiquarks pass close to each other. Occasionally a W or Z boson is produced by an energetic quark–antiquark collision but the other constituents usually suffer only a glancing collision. Consequently the other particles generally travel close to the beam directions and rarely have a large momentum perpendicular or transverse to the beams. In most collisions between antiprotons and protons, the constituents only experience glancing collisions because the quarks are very small and a direct collision is unlikely. As the collision energy is increased more particles tend to be produced and these have higher momentum in the beam directions. However, the average transverse momentum of these particles is usually less than 1 GeV even at very high beam energies. This result is of great significance when we are trying to detect the decay of W and Z bosons.

We are concentrating here on the electron decay modes of the W and Z bosons as these were used in the first searches for these particles. The expected masses of the W and Z bosons are around 80 and 90 GeV respectively. The decay $Z^0 \rightarrow e^- e^+$ involves decay products with a total mass of only 1 MeV! The decay $W^- \rightarrow e^- \bar{\nu}_e$ includes the massless antineutrino and so this leaves an even smaller mass in the decay products. We will consider here the decay of the Z^0 to electron–positron but clearly the W decays will result in decay particles of very similar energies.

If the Z boson is produced at rest, then from the conservation of momentum its decay products cannot have any net momentum. Consequently the electron and positron must be emitted back to back. As the masses of the electron and positron are negligible the mass energy of the Z boson must be converted into the motion energy of the two decay particles. As their momenta have to balance, the electron and positron must always carry 45 GeV of energy but their decay angle to the beam can vary. If this angle is very small, then the Z boson decay particles will be emitted along the beam with the beam fragments, and will be impossible to detect. If the decay occurs perpendicularly to the beam, then the electron and positron will be each carrying 45 GeV of transverse momentum and this is a truly characteristic signal. For most of the intermediate angles of decay, the signal is still very striking, as a large component of the 45 GeV momentum will be in the transverse plane and this will be much larger than the typical 1 GeV carried by the majority of the particles.

In most of these decays both the electron and the positron will have an outstandingly large transverse momentum. How can we prove that such events are really indicating the presence of a Z boson? The transverse momentum of both electron and positron have to be large in the same event and roughly to balance each other but how can we check that these energetic particles are coming from an object with mass close to 90 GeV? We can use the same technique that we met earlier when we introduced the discovery of short-lived resonances. If two particles are produced from the decay of a single parent then we can use energy and momentum conservation to deduce the energy and momentum of the parent. As always we have to combine the momenta by a method which includes the direction in the calculation.

Once we have the energy E and momentum P of the parent, its mass M can be evaluated from the formula $E^2 = P^2 + M^2$. Let us assume that the momenta (P) of both the electron and the positron were very accurately measured and that many examples of this type of Z^0 decay were observed. The masses (M) of the electron and positron are known and so their energies (E) can be calculated. Consequently the energy and momentum of the parent Z^0 boson can be deduced and its mass calculated. The masses of these Z events should all be near 90 GeV but we would not expect the masses to all be identical because of the uncertainty principle. The Z^0 lifetime is so short that its mass is expected to vary by a few GeV about its central value.

The decay of W^- to electron–antineutrino should behave in a similar way to the Z^0 decay we have just discussed. The W^- is slightly lighter at 80 GeV and so its decay products will each carry 40 GeV and the measurements will be complicated by the presence of an antineutrino! However, the electron carrying a large transverse momentum will be the signal which will be used in the search for the W boson.

6.6 How often can they be produced?

We have now considered the decay of W and Z bosons and discussed how often they are predicted to decay into various particles. We noted that their creation requires enormous energy and that this can be provided by a head-on collision between a proton and an antiproton. We must now try to estimate the number of W and Z bosons that will be produced when beams of protons and antiprotons collide. We know that the proton is composed of three valence quarks, two up and one down quark which carry around one half of its momentum. The rest of its momentum is carried by gluons and a large number of virtual quark–antiquark pairs.

When a W or Z boson is created, it is produced from the collision of a quark and an antiquark. We have seen that we need collision energies of around 80 to 90 GeV to produce these heavy objects. The proton and antiproton beams can each have energies of up to several hundred GeV at CERN, and this at first sight seems to be quite sufficient. However, we must allow for the fact that each valence quark in the proton only carries on average one-sixth of its momentum. This resulting average energy of 50 GeV per beam for the colliding quark and antiquark is only barely enough. Sometimes we will be lucky and get an interaction between quark and antiquark which carry more than their average share of energy, but this will not happen very often.

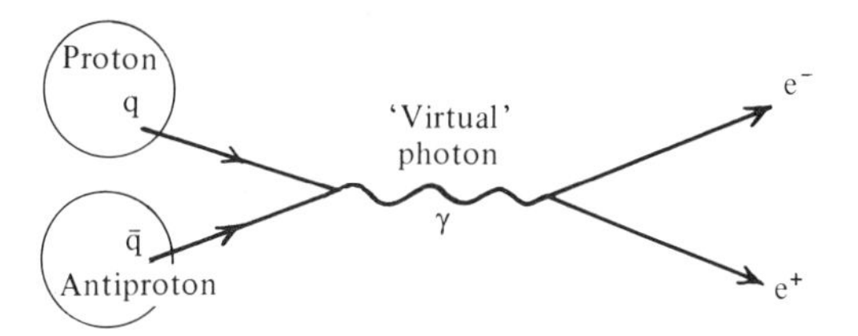

Figure 6.3. A collision between a proton and an antiproton produces an electron pair from the annihilation of a quark and an antiquark ($q, \bar{q}$). This Drell Yan process is very similar to the production of the Z^0 boson as that also involves the annihilation of quark and antiquark. As always there will be other particles produced in the violent collision.

Over the past decade a process which is very similar to the production of W and Z bosons has been extensively studied. In this so-called Drell Yan process two hadrons collide together and a constituent quark and antiquark annihilate to produce a virtual photon which then converts into an electron–positron or muon–antimuon pair (figure 6.3). When a W or Z boson is created the production mechanism is very similar (figures 5.3 and 5.4). In each case a single quark and antiquark pair annihilate to produce the massive boson, which then decays to its observable decay products. The rest of the proton and antiproton are not strongly involved in these production processes and continue more or less unaffected. This provides a background of extra particles to each collision but these are usually clearly distinguishable from the W or Z decay products.

As the theoretical predictions of the Dell Yan process agreed well with the experimental data, there was optimism that a similar calculation for the production of W and Z bosons would be reliable. The calculations for the production of a W boson in antiproton–proton collisions at a collision energy of 540 GeV were made and this was compared to the total number of collisions to be expected at this energy. If the calculations were right it would require over 10 million antiproton–proton collisions before an observable decay of the W boson could be expected!

This emphasises that the production of an energetic collision is just the first step in a very difficult experiment. The Z boson was predicted to be produced in an observable way, around 10 times less frequently than the W boson. A very selective method has to be found to extract the rare W and Z boson events from an enormous sample of other collisions. This has to be done without applying selections which are too severe, as these might invalidate the whole selection procedure.

The total number of antiproton–proton collisions that can be observed in an experiment depends on the beam intensities and sizes. Later we will see that antiproton–proton collisions were studied with collision rates as high as 5000 interactions per second and this poses another major problem. It is not possible with current technology to record anything like this frequency of events, as an enormous amount of information has to be recorded from a complex apparatus. As all events cannot be recorded most events have to be ignored, so how can we be sure that the very rare W and Z boson events will be retained? The W and Z events are only expected in one ten-millionth of the collisions and so we can ill afford to miss any of them!

It was clear that the accelerators and detectors would need to work reliably for a long time to have a chance of detecting a W or Z boson. Even then it was only expected that a handful of W or Z bosons would be recorded in several months of data taking. It was essential that all parts of the experiment ran efficiently, to avoid the risk of missing these rare events.

6.6 Can they be identified?

What else can we learn about the way that W and Z particles will be produced? Consider again the quark content of the colliding beams where a fraction of the proton's momentum is carried by a quark. On average this fraction will be one-sixth because of the presence of three valence quarks and gluons which carry half the momentum. An identical argument applies to the valence antiquarks in the antiproton. These quarks and antiquarks carry large amounts of momentum along the beam direction but very little in the transverse direction, which is perpendicular to the beams. When a W or Z boson is produced from a quark–antiquark collision (figures 5.3, 5.4) the boson will not have any significant transverse momentum as the colliding quark and antiquark move along the beam direction. However, the momentum of the quark along the beam is very unlikely to balance exactly that of the antiquark and so the boson will usually be produced with some momentum along the beam direction.

We discussed the decay of the W and Z boson in an earlier section, under the assumption that it was produced at rest. In this case the two decay particles are exactly back to back in space and carry precisely the same energy. How is this changed if the boson is produced with some momentum along the beam direction? This motion forces the two decay particles to be at an angle to each other and to carry somewhat different energies. However, the decay particles are still very energetic and their transverse momenta are

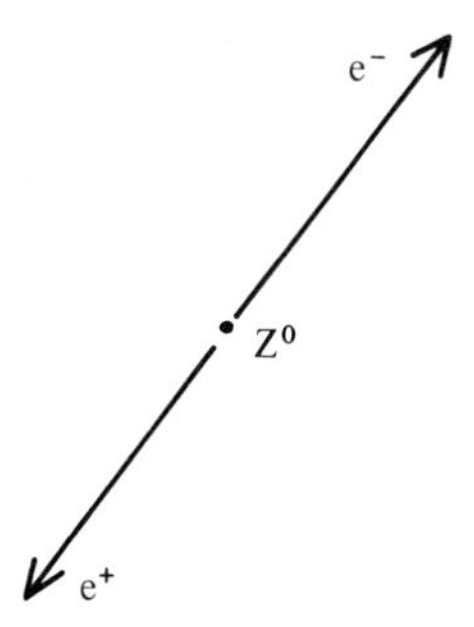

Figure 6.4. The decay of a Z^0 boson into an electron and a positron when viewed along the direction of the colliding antiproton–proton beams. The Z^0 boson is expected to be produced with little sideways motion and so the back to back decay is essential for momentum conservation. The decay products have to carry away all the Z^0 mass energy and so they are both emitted with very high energy.

unaffected by this motion of the boson along the beam. One very simple property does remain. When viewed from along the beam direction, the decay products of the bosons should still be back to back, which is a characteristic feature of such events (figure 6.4). Are the W and Z bosons expected to decay equally into all angular directions? The predictions here, especially for the W boson, are very interesting but we will return to this later in the story.

6.7 Summary

We have now introduced all of the important features of the W and Z bosons. This chapter has concentrated on various aspects of their detection without getting involved in the accelerators and particle detectors that will be described in the following chapters. We have seen that enormous energies are required to produce these particles and that they only exist for a tiny fraction of a second before decaying into more familiar particles such as the electron. The procedure used to identify such a rapidly decaying object is necessarily indirect. However, we have seen that measurement of the decay products can be used to reconstruct the mass of the short-lived W or Z boson. In the decay of the W boson, one of the decay products is an antineutrino which escapes without detection. We have postponed until later the details of how the energy of this elusive particle can be inferred from the experimental detector.

The decay products are recognisable because of their large transverse momenta, but there are still formidable problems to overcome. The W and Z bosons are produced only very rarely, even in very high energy collisions. Somehow, most of the more common interactions have to be suppressed and yet all the collisions producing W or Z bosons need to be recorded without serious loss. The expected number of W and Z events is very small and so every effort will be made to observe and record them all. Fortunately other types of reaction can be used to help make the prediction of W and Z boson production rates fairly reliable. The detailed knowledge of the inner structure of the proton, which has been built up from many earlier experiments, is a crucial ingredient to these predictions.

The scene is set for the experimental search. The theoretical predictions are unusually definite and precise. With efficient and reliable operation of accelerators and detectors, over a continuous data-taking period of several months, a few W and Z bosons should be detectable if they really exist. We have now discussed what was expected – but what really happened? We now have to introduce the accelerators, particle detectors and people involved in this search. Then we can follow the experimental search for the W and Z bosons.

Part Three

The tools for the search

Chapter Seven

Accelerators and CERN

7.1 What are accelerators?

The energetic cosmic rays which continually bombard the surface of the earth provide a natural accelerator for the study of high-energy collisions. Unfortunately the number of really energetic cosmic rays is far too small to allow the study of anything more than very common interactions. It is also very difficult to interpret the collisions as the exact nature of the cosmic ray is frequently unknown. For this reason artificial accelerators are also required.

Accelerators are used to produce controlled energetic collisions between subatomic particles. Some of the kinetic energy of the beams is used in the creation of additional particles, which frequently are unstable and only exist for a small fraction of a second. Detectors at the accelerator measure the properties of these new particles and, as we have seen, only particles with certain masses are found to exist. The classification of these particles has led to a deeper understanding of the proton and has produced the first evidence for quarks. The accelerator provides an intense beam of highly energetic particles which are usually protons but electrons and other particles can also be accelerated. The use of controlled intense beams allows the careful study of very rare interactions and these have frequently revealed important new discoveries. As the beam energy is increased, more energy is made available for particle creation and consequently heavier particles can be produced.

At the source of a proton accelerator, atoms are ionised into protons and electrons. The protons are accelerated to high speeds by the repeated application of accelerating electric fields while travelling through a region of magnetic field, which maintains them in a circular orbit. In a few seconds, the acceleration cycle has finished and the protons are travelling at nearly the speed of light, when they can be ejected from the accelerator and used to study high-energy collisions with a variety of targets.

How do accelerators work?

All of the accelerators that we shall discuss use electric and magnetic fields for their successful operation. Although the strength of the fields used in particle physics are high they are the same in nature as those observed in many day to day activities. The electric battery is a familiar example of an object which can produce an electric field. Consider an arrangement where each terminal of a battery is connected to a separate metal plate. This will produce an electric potential difference of a few volts between the plates, and a charged particle will be attracted towards one of the plates. This attraction will have the same strength and direction for both stationary and moving charges in this region.

The earth has a magnetic field of a few gauss, which has always played an important role in navigation because of the use of the magnetic compass. We are also familiar with bar and horseshoe magnets but the most powerful magnets are electromagnets, where an electrical current is passed through a coil of wire to produce a magnetic field. In a magnetic field a stationary charged particle feels no force at all but if it is moving then it experiences a sideways force which is perpendicular to its direction of motion and also to the magnetic field. The size of this force just depends on the charge of the particle, the magnetic field strength and the velocity of the particle. The action of a sideways force continually modifies the direction of a charged particle and it travels in a circle in a magnetic field.

These effects of electric and magnetic fields on moving charged particles are the essential ingredients to all of the accelerators we shall be discussing. Each accelerator uses electric fields to accelerate the charged particles to high velocities and magnetic fields to limit their motion to circular orbits. These fields need to be very strong because the charged particles are frequently travelling at very high velocities; in fact the accelerating electric fields can be as large as a million volts and the magnetic fields often exceed several thousand gauss.

What is accelerated?

For the study of the atomic nucleus, particles need to be accelerated to energies of millions of electronvolts (MeV). As already stated this is the energy gained by an electron, or particle of the same charge, when accelerated through an electric potential difference of one million volts. For the exploration of the nucleus and its interior we need very large batteries!

The accelerators that we shall be discussing all accelerate protons which are contained in each atom, so there is never any shortage of protons to be accelerated. Consider an atom of the simplest element, hydrogen, which contains just one proton and electron. The atom is electrically neutral and cannot be accelerated by electric fields. The electron has to be ejected from the atom in a process called ionisation, leaving the positively charged proton which can then be accelerated. This ionisation is often achieved by passing X rays or other ionising radiation through a container of gaseous hydrogen and then a strong electric field is applied to separate the protons from the electrons.

The cyclotron

The cyclotron (figure 7.1) is an accelerator that is used to accelerate protons up to energies of five million electronvolts, which is high enough to allow the study of the nucleus. A cyclotron, which is several metres high, fits into an average-sized room and the whole circular area of the cyclotron, up to 1 metre across, is enclosed between the poles of a very powerful magnet. A source of protons

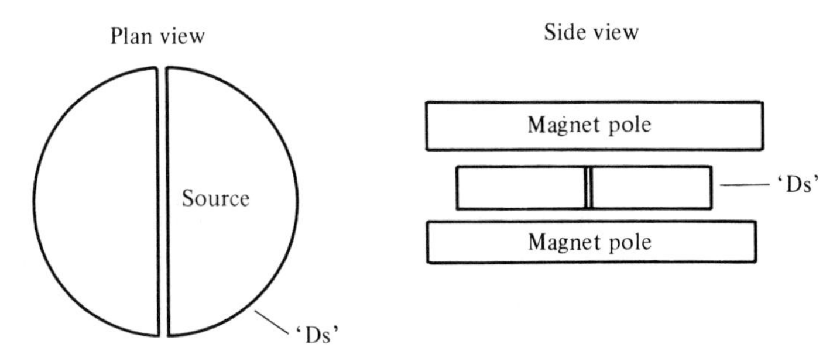

Figure 7.1. Two schematic views of a cyclotron. The side view shows that the 'Ds', which contain the accelerating particles, are enclosed in a uniform vertical magnetic field. The source of charged particles, proton or other positive ions is located at the centre of the cyclotron. The 'Ds' are electrically isolated from each other and held at different potentials. The particles move in a circular orbit, their energy being increased at each crossing between Ds by the accelerating voltage, which is continuously changed in sign at very high frequency. As the particles gain in energy the orbit reaches the edge of the cyclotron and the beam is ejected towards an experiment.

is placed at the centre of the cyclotron and these are first accelerated by a strong electric field. As the protons are in a constant magnetic field, they travel in a circular orbit with a radius which depends on their energy. The two halves of the cyclotron, called Ds because of their shape, are held at different electric potentials and as the protons pass through the gap between the two Ds they are accelerated by applying voltages of the appropriate sign. As they come back through the gap to complete the first circuit, the electrical voltage has to be changed so that they are again accelerated. Conveniently, as the velocity increases, the radius of orbit increases in proportion and so it always takes the same time for a particle to make one revolution. This means that the voltages between the Ds can be reversed at a constant rate which is typically one million times per second, or radio frequency. The energy of the particle is gradually increased by these small accelerations over thousands of subsequent rotations, and eventually it becomes so large that the orbit reaches the outer limits of the magnetic field. The particles are then extracted by an electric field and transported along a beam line to an experiment.

The operation of a cyclotron is restricted to relatively low energy because of the problems with maintaining uniform magnetic fields over large areas. In addition, extra complications are introduced to the required frequency of the voltage across the Ds, from the relativistic increase in mass of the beam particles.

Simple beam line

In order to steer a beam to a detector more magnets are necessary and the particles are conveyed along an evacuated beam pipe, through several bending magnets and made incident on a target. For studies of the nucleus this target is usually enclosed as part of a scattering chamber which incorporates detectors, a few centimetres in size, to study the reaction products of beam target collisions. As well as bending magnets there are also quadrupole magnets in the beam line. These each have two north and south poles (figure 7.2) and are used to focus the beam of charged particles in a similar way to the focussing of a beam of light by an optical lens. Particles on the axis of a quadrupole are unaffected, but off-axis particles do experience a sideways force. In fact, an off-axis particle will always be deflected back towards the axis in one plane and deflected away from it in the other. In other words a quadrupole magnet always acts to focus the beam in one direction and defocus it in the other, so what have we gained? If we install a second quadrupole with its poles

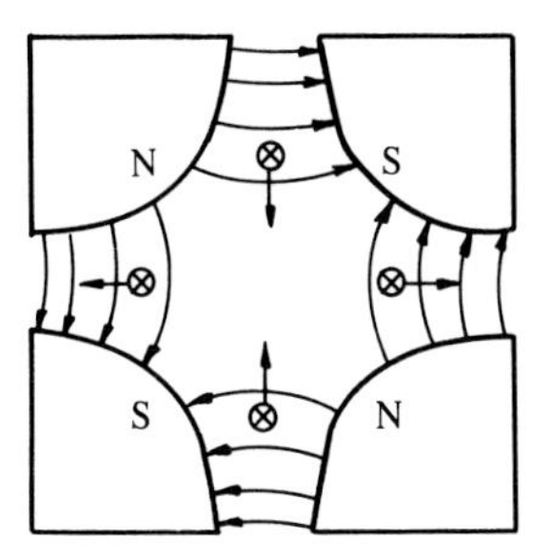

Figure 7.2. The magnetic quadrupole can be used to focus particle beams in a similar way to the focussing of a beam of light by a lens. Consider a positively charged particle travelling into the page near the centre of the magnetic quadrupole. The direction of the force on the particle can be deduced from the directions of the magnetic field lines and the particle's motion. A single quadrupole always has a net focussing effect in one plane but a defocussing effect in the other as shown. However, a pair of these elements produces a net focussing in both planes.

reversed, then this will focus in the opposite plane and this quadrupole doublet does indeed focus the beam in both directions. At first sight this seems strange but careful study confirms that an overall focussing effect is expected. These quadrupole doublets provide the beam focussing in even our largest accelerators and make it possible to transfer such beams over large distances without major beam losses. The larger particle accelerators usually have many beam lines, which transport accelerated particles to many separate experimental detectors.

Synchrocyclotron

Protons can be accelerated up to 750 MeV by using a synchrocyclotron. This uses the same principle as the cyclotron but includes one important extra feature. As a particle increases in speed, its mass increases and it gradually takes longer to make a circuit of the accelerator. Consequently it falls out of step with the accelerating field which is applied between the Ds. So the radio frequency electric field, which gives the accelerating voltage, is changed to a variable frequency to keep in step with the particles. On each acceleration, the radio frequency could typically drop from 20 to 10 million cycles per second which is a very clear illustration of the measurable effects of relativity in the laboratory. This whole acceleration process is repeated up to 50 times each second so we have to get used to considering very small distances and very small times. In a fleeting second, 50 separate accelerations of protons to hundreds of million electronvolts takes place.

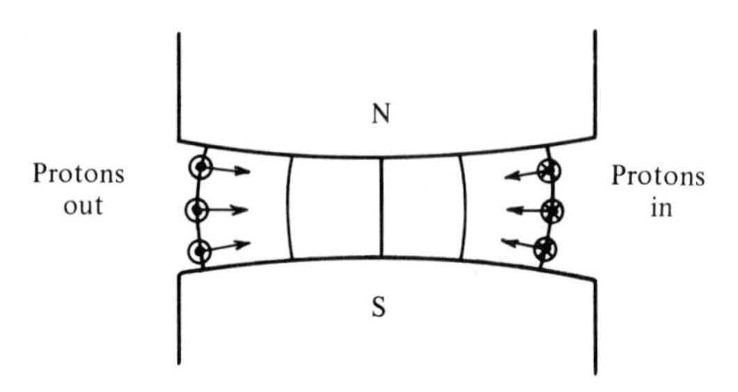

Figure 7.3. The tapered magnets of a weak-focussing accelerator produce a net focussing vertically because of the curvature of the magnetic field lines towards the edge of the magnet.

The synchrocyclotrons introduce another new feature: their magnetic pole faces are tapered. As the magnetic field is not exactly vertical towards the edges of a tapered magnet, this produces a vertical focussing of the beam as shown in figure 7.3. The largest synchrocyclotron in the world is over 4 metres in diameter, weighs 4000 tonnes and was constructed at the Lawrence Berkeley laboratory in California and was converted from a cyclotron in 1949. To double the size of such an accelerator requires an increase of a factor of eight in its weight and volume and so different techniques are used to achieve even higher energies.

Synchrotron

For studying the properties of particles smaller than the nucleus energies of thousands of millions of electronvolts are required (GeV). The synchrotron is the most widely used accelerator to achieve energies in this range. As all the accelerators used in the proton–antiproton collider project were synchrotrons then we will discuss them in a little more detail. First we will outline the main concepts, as there are several new features to discuss, and then discuss the operation of a synchroton.

The main problem in extending the synchrocyclotron to even higher energies was the size of the magnet that would be necessary. In the synchrotron this problem is overcome by using a ring of magnets arranged at the outer edge of a circle, rather than a huge magnet covering the whole region. To enable particles of all energies to travel in the same orbit, it is then necessary to vary the magnetic field during the acceleration. The original beam can no longer spiral out from the centre and has to be injected into the outer ring directly from some smaller accelerator frequently at energies of several hundred million electronvolts. Radio frequency cavities provide electric fields at intervals around the circumference and their frequency has to be gradually increased to keep in step with the accelerating particles. The particles are localised in a bunch and so the accelerating electric fields are switched on as the bunch enters the cavity and reversed before they leave. This means the protons are pulled into the accelerating region and then pushed out, so two separate bursts of acceleration are obtained.

When a beam is injected into an accelerator each particle will have a slightly different path through the machine and as they travel along slightly different routes they will experience different magnetic and electric fields. These differences may be small, but the effect will be that each particle will not travel exactly around an ideal orbit. Instead each particle will follow a slightly different path and move above and below and inside and outside the ideal orbit in an oscillatory motion. If there were no beam focussing, these oscillations would gradually destroy the beam as the particles would hit the magnets or outside of the beam pipe. The earliest synchrotrons used magnets with tapered poles which focussed the beams weakly in both horizontal and vertical planes. These beams were maintained but very large vacuum chambers were needed to contain the beam and this required expensive large-aperture magnets. One of the largest weak-focussing synchrotrons was constructed at the Dubna laboratory in the USSR. This comprises magnets weighing over 30000 tonnes and can accelerate protons to 10000 million electronvolts (10 GeV). It also has a very large beam pipe with a cross-section of 1.5 by 0.4 metres and the correspondingly large magnet is the main reason for its very large weight.

The final step to the currently used strong-focussing synchrotrons was an important one. The strong focussing was achieved by using magnets with tapered poles as before but this time they were arranged such that every second magnet had the taper in the reverse direction (figure 7.4). This pair of magnets acts like a quadrupole and provides strong focussing both vertically and horizontally. This means that the beam oscillations about a stable orbit are compressed and so the beam pipe and the aperture of the magnets can be reduced to a radius of only 10 centimetres! In some synchrotrons the same effect is achieved by using flat magnets together with pairs of quadrupole magnets around the ring. It is now possible with this technique to achieve energies of one thousand, thousand million electronvolts (1 TeV) as at the Fermi National Accelerator Laboratory (FNAL) which is near Chicago in the USA. All the high-energy synchrotrons currently used at CERN, near Geneva in Switzerland, also use the strong-focussing technique.

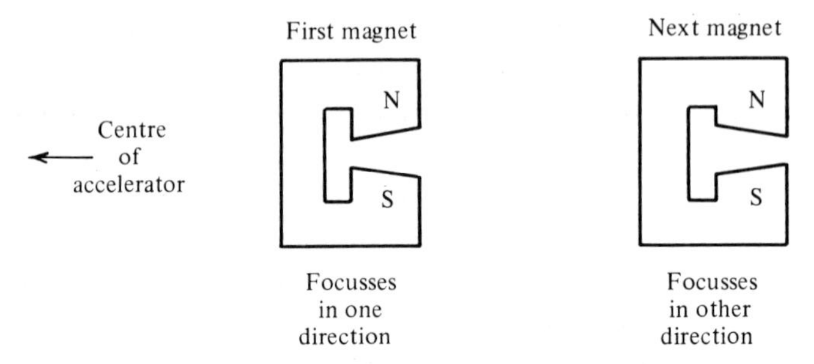

Figure 7.4. In a strong-focussing accelerator the tapering of the magnets is reversed for each successive magnet. The focussing is achieved by the curved magnetic field lines and the variation in magnetic field strength across the magnet. These accelerators are much more economical because the magnets can be made with a much smaller aperture when the beam is more localised. Nowadays this is often achieved by using a mixture of magnetic dipoles and quadrupoles rather than tapered magnets.

7.2 What is CERN? A brief history

The CERN laboratory, which is situated on the outskirts of Geneva, straddles the border between France and Switzerland. It now has an unequalled range of research facilities, including a wide range of accelerators, for the investigation of the structure of matter and the fundamental forces. There are 13 member states – Austria, Belgium, Denmark, West Germany, France, Greece, Italy, Netherlands, Norway, Spain, Sweden, Switzerland and the United Kingdom. These countries contribute to the budget of the organisation in proportion to their net national revenue. CERN employs about 3500 people, including around 1200 scientists and engineers, and also provides research facilities for over 2500 physicists from the member states. Other visiting scientists come from many other countries including the United States, the Soviet Union and the People's Republic of China. CERN is one of the most impressive examples of international cooperation in science and currently has an annual budget of 680 million Swiss francs, which is equivalent to 220 million pounds at 1984 exchange rates. This budget approximately corresponds to a contribution of fifty pence per year from each inhabitant of the member states.

It is primarily a laboratory for pure research but there are frequently applications of the techniques developed for particle physics in other areas of science, medicine and industry. Much of the equipment that is required involves the most up to date technology. None of the research work at CERN is secret and the experimental results are published in scientific journals which are available in libraries throughout the world. No military or nuclear power research is carried out at CERN. It is devoted to the exploration of the inner structure of the atom and the understanding of basic forces.

What are the origins of this successful laboratory? The idea of a European Laboratory was first raised publicly by the French physicist de Broglie at the European Cultural Conference in Lausanne in 1949. It was becoming clear that the accelerator projects that were needed to make progress in studies of the nucleus were beyond the resources of individual European nations. Under the influence of UNESCO, meetings were held in 1950 and these led to the establishment of an international nuclear physics laboratory. In February 1952 eleven governments signed an agreement which set up a provisional 'Conseil Européen pour la Recherche Nucléaire' which was the origin of the currently used abbreviation of CERN. Later that year a site for the laboratory near Geneva was offered and in the following year a referendum in the canton of Geneva supported this proposal. The convention establishing CERN was signed, subject to ratification, in Paris on 1 July 1953. This stated the aims – 'The Organisation shall provide collaboration among European States in nuclear research of pure scientific and fundamental character, and in research essentially related thereto. The Organisation shall have no concern with work for military requirements and the results of its experimental and theoretical work shall be published or otherwise made generally available.'

From the beginning it was decided to build a high-energy proton synchrotron (PS) using the latest ideas from accelerator technology. To gain experience for this ambitious venture a smaller, less powerful synchrocyclotron was also constructed. The excavations were started on the Geneva site in May 1954 for the construction of a 25 GeV PS, a 600 MeV synchrocyclotron, together with their associated experimental areas and other accommodation. Gradually, the first buildings appeared, the first proton beam was circulating in the synchrocyclotron on 1 August 1957, and during the following year the first direct observation of the decay of a pion to an electron and an antineutrino was made at this accelerator.

The large PS required 100 30-tonne magnets to be very accurately located in a ring of 200 metres diameter and this magnet installation was completed on the 10 July 1959. At the same time the construction of a bubble chamber and other electronic detectors was being completed. The PS first accelerated protons to a world record energy of 24 GeV on the 24 November 1959. It was the first machine to operate with strong focussing of the protons and its energy made it the most powerful accelerator in the world. The 10 GeV Dubna accelerator in the USSR was the previous energy record holder and a bottle of vodka had been sent from Dubna to CERN as a contribution to the celebration. The empty bottle was returned containing a polaroid photograph of the oscilloscope trace of the 24 GeV acceleration!

In the early 1960s the first neutrino beams were produced and the first ever bubble chamber picture of a neutrino interaction was recorded in 1963. As the more powerful accelerator became fully utilised the detectors and computers used to analyse the particle collisions and decays became much more sophisticated. A series of hydrogen bubble chambers were used for the analysis of particle collisions, including the CERN 2 metre chamber which started operating in December 1964. Millions of pictures were taken with these chambers and these were analysed in laboratories throughout the world, leading to the discovery of many short-lived particles and a better understanding of the interactions between them.

In 1965 a lease was signed with France to allow the original site to be extended over the Switzerland–France border. This was primarily to allow construction of the intersecting storage rings (ISR), two interlinked rings of 300 metres diameter which cross at eight points and which would allow head-on collisions between two proton beams each with energies up to 30 GeV. Protons were to be accelerated in the PS and then transferred and stored in the ISR for many days. On the 27 January 1971 the first evidence of proton–proton interactions in colliding beams was observed at this accelerator. The collision energy of 60 GeV was the highest achieved in the world until the introduction of the antiproton–proton colider at CERN in the 1980s. A 3.5 metre diameter, big European bubble chamber (BEBC) was constructed at CERN and became operational in 1973. This chamber contains over 40000 litres of liquid hydrogen which is kept at a temperature of −247 degrees centigrade. In the same year the first evidence for weak neutral currents, where a neutrino interacts and preserves its identity, was observed in a heavy-liquid bubble chamber called Gargamelle at CERN. This chamber was filled with freon (CF_3Br) which provides a larger target nucleus than hydrogen for the weakly interacting neutrinos.

In 1971 a project for the construction of a 300 GeV super proton synchrotron (SPS) was approved. This comprised 1000 magnets around a ring 2.2 kilometres in diameter and was to be located in a tunnel more than 20 metres under the ground. By 1974 the drilling of the 7-kilometre tunnel had been completed, and the drill had returned to its initial position within an accuracy of just 2 centimetres!

Over the next year 1000 SPS magnets were installed, the last one being finished in December 1975, and on the 17 June 1976 the SPS began operation at an energy of 400 GeV. The beam intensity of the SPS soon exceeded its design figure and in 1978 the peak energy was increased to 500 GeV. At that time it joined the machine at the FNAL (USA) as the highest-energy accelerator in the world. A British engineer, the late Sir John Adams, made crucial contributions to the success of the accelerators at CERN. He played a central role in the construction and operation of both the PS and the SPS. The SPS provides particle beams to two large experimental areas to the north and west of the CERN site, one in France and the other in Switzerland. A variety of beams including those made up of photons, electrons, muons, neutrinos and pions can be created from the original energetic protons by bombarding an external metal target with the protons. Electric and magnetic fields together with absorbers are used to separate the reaction products into well-defined beams of particles.

An aerial view of part of the CERN site is shown in figure 7.5*a*. An impression of the complexity of accelerator control and the scale of CERN's computing facilities can be obtained from the views of the SPS control room and the central computers which are shown in figure 7.5*b*, *c*. This very brief summary of the major events in the evolution of CERN brings us up to the preparations for the antiproton–proton collider project but we will be looking at this in detail later in this chapter.

To complete this survey of the accelerators at CERN we should mention again the large electron–positron collider (LEP) which is currently under construction. In its initial version, which should begin operating at the end of 1988, it will provide head-on collisions between beams of electrons and positrons each with energies of 50 GeV. These beams will be circulated in a tunnel of 27 kilometres circumference, which was started in 1983, adjacent to the existing CERN site. As discussed in chapter 5 the electron–positron collisions provide the 'cleanest' environment for studies of the Z^0 boson as this particle can be produced from the annihilation energy.

We have briefly described the major pieces of equipment and some of the important dates in the history at CERN where the collaboration between individuals from 13 member states really does work very well. These successful collaborations often include contributions from countries throughout the world.

Figure 7.5. (*a*) An aerial view of the main CERN site which is situated very close to the border between France and Switzerland on the outskirts of Geneva. The CERN laboratory is in the foreground and the three circles of increasing size indicate the PS, ISR and SPS accelerators. (*b*) A view of the main control room of the SPS which is manned around the clock. The highly complex antiproton–proton collider operation was controlled from here by the operations staff with the help of a number of computers and diagnostic displays. (*c*) A general view of the central computers at CERN which were an essential element in the rapid analysis of the collider data. These computers are accessed by over 750 terminals on the site and also by data communications networks from all over the world. Photographs courtesy of CERN.

(*b*)

(*c*)

7.3 Other major laboratories

How does the CERN laboratory compare with other particle physics research laboratories around the world? Over the years the accelerator with the highest available energy has at various times been in the USA, USSR and Europe. This competition has helped to maintain a vigorous programme of research work in the various countries and this process continues today. There are now many accelerators in the world which can accelerate particles to energies of a few GeV. Let us concentrate here only on those accelerators that are already operational at very high energies or that are planned to be in operation before the end of the decade. This leaves us with just four other laboratories, two in the USA, one in the Federal Republic of Germany and one in Japan. There are many other famous laboratories which have been very successful and still operate accelerators but these are no longer being operated at frontier energies.

The only other laboratory which has a very high energy proton synchroton is the FNAL, which is located about 50 kilometres south of Chicago. The accelerator magnets at this laboratory have recently been converted from conventional to superconducting magnets and this has enabled the energy to be increased from 500 GeV to one thousand GeV (1 TeV). A comprehensive programme of antiproton–proton collider physics is now under preparation at FNAL. This should start in 1986 and will involve collisions between beams of 1 TeV protons and antiprotons, which will then be the highest-energy collisions ever achieved at any accelerator. This laboratory provides the competition to CERN for the study of energetic antiproton–proton collisions.

A unique accelerator, known as HERA, which will collide a 800 GeV beam of protons with a 30 GeV electron beam is now being constructed at the DESY laboratory, which is in the outskirts of Hamburg in West Germany. This should be completed by 1990 and will then be the best accelerator in the world for the study of the inner structure of the proton. The much larger energy of the proton compared to the electron accelerator emphasises that it is much more difficult to maintain an electron beam at a high energy in a circular accelerator. Many important discoveries have been made at electron–positron colliders. The annihilation of the electron and positron provide a localised concentration of pure energy which can subsequently convert into many types of particles. This is an ideal environment for the search for heavier quarks or leptons, providing the energy of the collider can be made large enough. At the Stanford Linear Accelerator (SLAC) in California the electron–positron collider (PEP) can produce 40 GeV collisions and there is a similar collider with a maximum energy of 46 GeV (PETRA) at the DESY laboratory in Hamburg. In Japan an electron–positron collider with a peak energy of 60 GeV (TRISTAN) is now being constructed and is due to start operating in 1986.

Finally an exciting project is underway at the SLAC laboratory to convert its linear electron accelerator into a single pass collider facility (SPC) which will collide electron and positron beams at collision energies of up to 100 GeV. The first beams are expected at the end of 1986 when it will be the only linear accelerator used for colliding beam physics. At the SPC the beams of electrons and positrons are not stored at high energy as in most colliders. The electrons and positrons are both accelerated to 50 GeV, at slightly different times, using the same linear accelerator. The beams are then steered so that they travel in opposite directions around a loop of magnets and are made to collide head on. Even though this process is repeated every fraction of a second it still requires very compact beams to produce a useful interaction rate. However, this may well be the only way to achieve much higher electron–positron collision energies because of the large power losses in circular electron accelerators.

To summarise, CERN has at present a very competitive set of accelerators which will be further enhanced when the LEP is constructed towards the end of the decade. However, there are several other laboratories which also have first-rate facilities. Results from CERN and the other laboratories have yielded major progress in the understanding of the structure of matter, and there is every indication that this rapid progress can be maintained in the next decade.

7.4 How are protons accelerated to 450 GeV?

In this story the CERN accelerators were used to produce high-energy collisions between protons and antiprotons. Before we leave the existing accelerators and start on our story, we will first describe how protons are accelerated to 450 GeV at CERN. When we are familiar with this procedure we can then examine the changes needed to convert the existing accelerators into an antiproton–proton collider. The layout of the CERN accelerators and their approximate relative sizes are shown in figure 7.6.

We have already seen that the 28 GeV PS was completed in 1959 and acted as a primary accelerator at CERN for many years. Now over 25 years later this same accelerator is the key source of protons for most of the other newer aceclerators. It has been modernised and enhanced but it is remarkable that it still plays such a central role in the CERN accelerator complex.

The SPS was constructed at a depth of more than 20 metres and a portion of the 6 metre diameter SPS tunnel is shown in figure 7.7. The protons are confined to an evacuated beam pipe which passes through the other accelerator elements that we have discussed, the bending magnets, quadrupoles and accelerating radio frequency cavities, and most of the air is pumped out of the steel beam pipe in order to reduce the number of collisions between protons and air molecules.

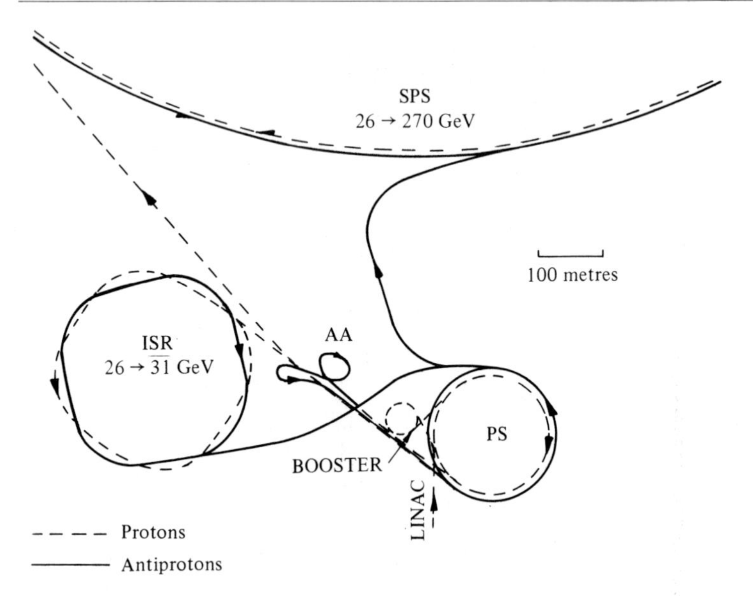

Figure 7.6. A schematic view of the flow of protons and antiprotons at the CERN accelerator complex for collider operation. Protons are accelerated in the linear accelerator (LINAC) and the booster before being injected into the proton synchrotron (PS) to be accelerated to 26 GeV. At this energy they are ejected towards the antiproton accumulator (AA), where they strike a target and produce secondary particles. From these secondary particles, antiprotons with energy close to 3.5 GeV are focussed and injected into the AA. This operation is repeated every 2.4 seconds for up to a day to collect sufficient antiprotons and these are 'cooled' during this process. For collider operation antiprotons are ejected from the AA and accelerated in the PS to an energy of 26 GeV before being injected into the super proton synchrotron (SPS). Shortly afterwards protons are accelerated to 26 GeV in the PS and injected into the SPS in the opposite direction to the antiprotons. Finally the SPS simultaneously accelerates the protons and antiprotons to 270 GeV and keeps them at this energy for up to two days. The intersecting storage rings (ISR), which are normally used to study proton–proton collisions, were also used to study antiproton–proton collisions up to energies of 60 GeV.

Protons are obtained from atoms by the ionisation of gaseous hydrogen and are first accelerated to an energy of 50 MeV in a small linear accelerator (LINAC). Each proton only has a minute charge of 1.6×10^{-19} coulombs, but the beam contains up to 10^{14} protons and as the beam is provided in just one ten-thousandth of a second the current carried by the proton beam is more than a tenth of an ampere. In order to increase the beam intensity, the protons are sent into a small four-ring synchrotron (BOOSTER) and accelerated to 800 MeV before being sent into the PS. The protons are accelerated in the PS up to 26 GeV and then transferred to the SPS for final

Figure 7.7. A view of the super proton synchrotron accelerator tunnel which is 20 metres below the surface. The ring of magnets and focussing quadrupoles are distributed around a circle with a diameter of 2.2 kilometres. At intervals around the circumference, radio frequency accelerating fields are applied to increase the energy of the beam particles. In just a few seconds protons are accelerated to a speed close to that of light. Photograph courtesy of CERN.

acceleration to 450 GeV. In normal operations this cycle is repeated every 10 seconds day and night.

It is clear that even the conventional operation of these accelerators is extremely complex, and requires sophisticated monitoring equipment and experienced personnel. Many experiments have been successfully completed at CERN by scientists and engineers from many different countries. The reliable operation of the accelerators has been a crucial ingredient to the success of these experiments.

7.5 Antiprotons. Production, separation and storage

How do we obtain antiprotons?

The PS in normal operation can accelerate protons to 26 GeV every 2.4 seconds. When these protons collide with a metal target some of this energy is converted into mass and extra particles are produced. These are created with a very wide range of energies and at many different angles and very occasionally when the proton interacts, it produces an antiproton amongst the other particles. When we measure the various momenta carried by these antiprotons we find that the most popular value is close to 3.5 GeV and even these antiprotons are produced only once in every million proton collisions. Although this is a very small yield, the PS can deliver around 10 million million protons (10^{13}) onto the target and so we might expect to be able to produce around 10 million of these antiprotons (10^{7}) every 2.4 seconds.

How can we separate the antiprotons?

The antiprotons can be separated from the other types of particles by using both electric and magnetic fields. We have seen that a magnetic field deflects oppositely charged particles in opposite directions and this can be used to select negative particles. The amount of bending in a magnet depends on the momentum of a particle and so the 3.5 GeV particles can be selected by choosing only those particles which come out deviated by a certain angle. Once we have selected a beam of negative particles, all with similar momenta, it will still include pions, kaons, antiprotons and other negatively charged particles. A feature that helps in the separation process is the relative stability of the antiproton, as most of the other particles decay very rapidly. Two particles with the same momentum but with different masses clearly travel with different velocities with the heavier particle travelling more slowly. If a mixed beam of particles is passed through an electric field, which acts perpendicularly to the beam, then the heavier particles are more deflected as they spend longer in the deviating field and this technique can also be used to separate antiprotons and other particles.

How can we store the antiprotons?

Antiparticles annihilate with particles and convert their combined mass energy into new particles and so our storage system must keep the antiprotons away from any matter. One obvious way to store the antiprotons is to use a ring of magnets so that the antiprotons can be kept in a stable circular orbit. As more antiprotons are produced every 2.4 seconds they can be added into the antiproton stack which is continually circulating with a momentum of 3.5 GeV. Imagine for some reason that we fail in this task, will we then have a major safety problem? Although we are studying the very highest energy collisions between particles, the total energy involved is very small because so few particles are involved. The energy of 1 GeV, which is very large on the atomic scale, corresponds to only 1.6×10^{-10} joules! Consequently even collisions of millions of high-energy particles only release energies of a small fraction of a joule, which is a very small energy by our normal standards.

7.6 Production of an intense beam of antiprotons

The antiproton beam with eventually need to have high intensity and be well focussed. As the number of antiprotons is increased, there will clearly be a mutual repulsion between the particles, which are all negative. At this stage the most difficult problem of the whole project has to be faced. Is it really possible to collect enough antiprotons in a small space so that, when such a beam is in collision with a proton beam, there are enough interactions for the W and Z bosons to be produced in significant quantities? In the early days of the project this was the greatest unknown and clearly the project could not be started until there was some detailed studies of this problem.

Fortunately Simon van der Meer, one of the foremost accelerator physicists, was working at CERN and studying this problem. Even if antiprotons with momenta of around 3.5 GeV can be stored in a ring of magnets, these momenta need to be very similar and the beam needs to be compressed in size. When this momentum and size compression has been achieved the beam is described as cooled, even though its temperature is not changed. The method of beam cooling that was developed at CERN was called stochastic cooling. In stochastic cooling the exact centre of the beam is measured at one point in the ring of magnets and its deviation from an ideal orbit is calculated. The appropriate size and sign of a correcting electric field is calculated and this information is sent across the accelerator, so that the correcting fields can be applied when the measured particles reach a second point. In this way the centre of the beam can be kept at the optimal position and the same procedure can be used both vertically and horizontally. For some individual particles the correction will be in the wrong direction, but on average the beam is gradually more concentrated by this procedure. The cooling is called stochastic, which means random, because the random motion of the beam particles is reduced. The early tests of stochastic cooling at CERN were performed at the ISR and these showed that by using electronics operating at frequencies of thousands of millions of cycles per second, that this procedure did indeed work.

7.7 The ICE project

As stochastic cooling was a crucial element of the whole collider project, it was decided to make a very detailed test before proceeding. This benefited from the existence at CERN of apparatus, in the form of a small ring of high-precision magnets, which had been used in another major experiment which measured the magnetic properties of the muon with great precision. As this experiment was finished the same magnets were used for tests of beam cooling and the new project was called the initial cooling experiment (ICE). The conversion of this magnet system and its associated electronics was completed rapidly and the ICE ring received its first protons in December 1977.

The requirement of the cooling system was that the beam should be compressed in three ways, vertically, horizontally and in its momentum spread and it was essential that this cooling be achieved over a period of several hours. A decision on the future of the antiproton–proton collider project was due to be taken by the CERN council in June 1978. This meant that the first six months of tests on ICE were especially important, as it was crucial to demonstrate that the required rates of cooling could indeed be achieved. The results were encouraging and showed that the cooling rates first achieved at the ISR could be significantly improved.

These early cooling results were very impressive and were indeed essential to the next stage of the collider project. Once it had been proved that stochastic cooling was practicable, the collider project was prepared in great detail and presented to the CERN council for approval in June 1978. The cooling tests continued and gradually many of the remaining uncertainties were removed and by May 1979 simultaneous vertical, horizontal and momentum beam cooling had been achieved for the first time. The rapid rate of this cooling was also impressive, with the momentum spread being reduced by a factor of 5000 during every hour. Although most of the ICE tests were done with protons a limited number used antiprotons. These were enough to prove that the antiprotons were certainly stable for several hours, much longer than the previously measured upper limit of 100 microseconds. This was very important because in the collider project it would take around a day to collect the antiprotons and these would be circulating in the accelerator for several days.

7.8 The antiproton accumulator

Now we must examine how the antiprotons were collected and cooled in the real collider project. ICE was just a test of the ideas

Figure 7.8. A view of the antiproton accumulator where the precious antiprotons are stored and cooled. The connections across the centre of the ring of magnets carry the electrical signals from the beam sensors to the kicker magnets which are used to compress the beam. Photograph courtesy of CERN.

of stochastic cooling. The heart of the final system is the antiproton accumulator (AA) which serves in a joint role as a collector and subsequent cooling device for the antiprotons (figure 7.8). This is a ring of magnets one-quarter of the size of the PS and it includes the fast electronics needed to perform the stochastic cooling on the stored antiprotons. To achieve a high yield of antiprotons it is important that the AA should accept as wide an angular and momentum range of incoming antiprotons as possible. For this reason the vacuum chamber was chosen to have a very large width of 70 centimetres. The air was pumped out such that it was held at a high vacuum of 10^{-10} torr which minimises the loss of antiprotons from collisions with gas molecules.

The production, storing and cooling of antiprotons

The PS accelerates a beam of protons up to an energy of 26 GeV every 2.4 seconds. When the protons have been accelerated to 26 GeV in the PS, they are extracted and the beam is made incident on a metal target close to the antiproton accumulator. The antiprotons with momentum around 3.5 GeV are then extracted from the other secondary particles by magnetic fields in a device called a magnetic horn and are stored in the AA. For every million incident protons, around one such antiproton is produced, so by using the full beam of 10^{13} protons, around 10^{7} antiprotons can be generated every 2.4 seconds. As it takes around a day to build up the required number of antiprotons it is most important that these newly produced antiprotons should not disturb any stored antiprotons. To help solve this problem a magnetically operated shutter isolated the outer and inner parts of the wide beam pipe of the AA. When a new set of antiprotons is delivered they are deflected to travel in an orbit on the outer side of the vacuum chamber. Stochastic cooling is immediately applied to these antiprotons by measuring their spatial properties at pick up stations and using kicker magnets to 'cool' them. After 2 seconds this procedure reduces the spread in momentum by a factor of 10 and these new antiprotons can be added to those that are already accumulated in the stack. At this stage the shutter is lowered and radio frequency fields are used to move these antiprotons to an orbit on the inner side of the vacuum chamber, to join the accumulated stack of antiprotons.

This sequence of collection and stacking is repeated until the required number of antiprotons has been collected, with the antiprotons spread around the full circumference of the AA so that the mutual repulsion of the negative charges is minimised. Nevertheless it is not possible to continually store antiprotons because eventually mutual repulsion will prevent further reductions in size. Stochastic cooling is applied to the stacked beam and over several hours a dense core of particles begins to form towards the inside of the vacuum chamber. As the cooling is continued, this dense core continues to become more concentrated and eventually is used as the source for the antiproton bunches in the collider experiments (figure 7.9). The cooling rates in the first cooling experiments at the ISR yielded results of a few per cent reduction in size per hour but in the AA, compression factors of over 100 million per hour are required. This emphasises how many improvements were needed in this beam technology in just a few years.

The process of storing and cooling the antiprotons is technically difficult and time consuming. The loss of electrical power, which is fairly common during the thunderstorms of winter, means that the patiently collected antiprotons are immediately lost. If an error is made in the transfer of these antiprotons between accelerators then again the whole collection process has to be restarted. These are the main disadvantages of using antiprotons in the experiment but the big advantage is that a single ring of magnetic and electric fields can be used to accelerate simultaneously protons and antiprotons in

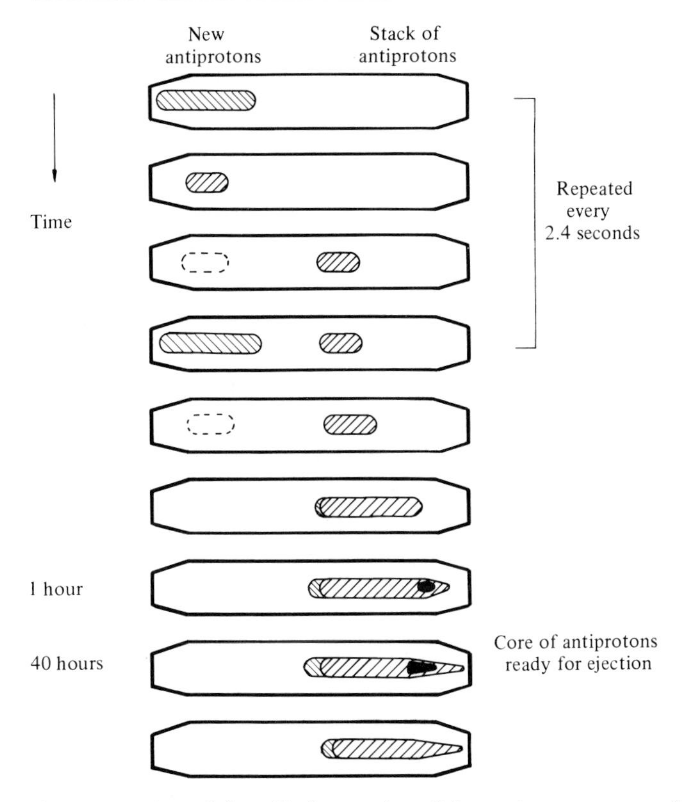

Figure 7.9. A cross-section of the wide beam pipe of the antiproton accumulator, illustrating the gradual compression of the antiproton beam. Every 2.4 seconds a new stack of antiprotons is injected into the outer part of the beam pipe and compressed by kicker magnets. After 2 seconds a magnetically operated metal shutter, which separates the inner and outer parts of the beam pipe, is lowered while the antiprotons are deflected into the inner part of the beam pipe. This process is repeated for each new pulse of antiprotons and the stack of antiprotons is also compressed by kicker magnets. Gradually a dense core of antiprotons is collected by this procedure and subsequently this is ejected into the super proton synchrotron for collider operation.

opposite directions. Any system that will accelerate protons will automatically accelerate antiprotons, provided they are travelling in the opposite direction. Consequently, the existing single ring SPS could be used to accelerate simultaneously protons and antiprotons.

7.9 The antiproton–proton collider project

The aim is to use the CERN SPS as a collider between protons and antiprotons at the highest possible energy. In normal operation the magnetic field is gradually increased to keep the beam of increasing energy at the same radius and the SPS can accelerate protons up to 450 GeV over a 10-second cycle. For the collider project, the magnets have to provide high magnetic fields for long continuous periods and this requires more magnet cooling and power. In collider operation the maximum field available from the magnets corresponds to an energy of 270 GeV for the beams.

How are the collisions of protons and antiprotons, each with beam energy of 270 GeV, brought about? We have covered all the separate elements and now we need to link them together. The flow of protons and antiprotons in the full set of accelerators is shown in figure 7.6. Consider first the antiprotons which are accumulated and cooled in the AA over a period of typically one day. Usually two-thirds of these 3.5 GeV antiprotons are ejected from the AA and sent around the loop of magnets towards the PS. In the PS the antiprotons are accelerated to 26 GeV and in the early days of the collider were delivered as a single bunch of antiprotons to the SPS. This 'bunch' is only a fraction of a millimetre across in the horizontal and vertical directions but it has a length of 30 centimetres. As it was very important to preserve the valuable antiprotons, these transfers were usually preceded by several rehearsals of the transfer procedure, with 'pilot shots' of a small number of antiprotons. When these showed that the transfer procedure was working well, a 'big shot' was attempted.

As the protons could be regenerated in a few seconds they were used wherever possible for tests of the system. The LINAC supplied protons to the BOOSTER and then the protons were accelerated to 26 GeV in the PS. At this energy they were ejected and transferred to the SPS shortly after the antiprotons, and injected in the opposite direction. When both protons and antiprotons are in the SPS they are both accelerated to 270 GeV and at this energy the acceleration is switched off and the magnetic fields are fixed to hold the beams at this energy. In later operations of the collider there were three separated bunches of protons and three of antiprotons circulating in opposite directions. These counter-rotating bunches collide at just six places around the circumference of the SPS. The bunches are already concentrated by the stochastic cooling but the collision probability can be further increased at the two intersections used for experimental measurements. Pairs of magnetic quadrupoles are used to focus the beams and in this way the collision rates can be boosted by almost a factor of 50.

To reduce beam losses from collisions with gas molecules, over 1000 extra pumps were installed to improve the vacuum of the SPS beam pipe. Also, the long-term stability of the magnets needs to be very high and so the stability of the magnet power supplies was also improved. There are small but significant losses at each stage of transfer or acceleration, and for the experiments to stand any chance of discovering the W and Z bosons, this whole complicated system of accelerators and transfers would need to operate very efficiently. Not only on a few occasions but on a regular basis, 24 hours a day, for several months!

Chapter Eight

Detectors

8.1 What needs to be detected?

The experimental search for the W and Z bosons has to take place underground in the super proton synchrotron (SPS) tunnel. The accelerator contains counter-rotating beams of 270 GeV protons and antiprotons which are each localised into 'bunches' of up to 10^{11} particles. These collide at the centre of a detector every 8 microseconds (8×10^{-6} seconds) and on average there will be an interaction in every 10 bunch crossings. However, the process is quite random, and so there will occasionally be interactions on successive crossings or even more than one in the same bunch crossing.

When an interaction occurs some of the 540 GeV collision energy is used to create new particles and in a typical interaction over 50 new particles are produced. The detector is required to record as much information as possible about these newly created particles, especially the energetic ones. Ideally it will allow the measurement of the momenta of all charged and neutral particles and also their identification. In practise this ideal is rarely achieved and some simple detectors record only a few important results of each collision.

At the antiproton–proton collider the two main detectors were designed to make comprehensive measurements of the collision fragments. The collision energy was 10 times higher than had been studied before and so it was very important that the detectors be able to record any unexpected features of these energetic collisions. However, they were also designed to detect the W and Z bosons and so the reliable measurement and identification of energetic electrons and muons was a crucial element in their design.

The operating conditions were very difficult as the interactions occur so rapidly after each other. In order to measure many details of each event it is necessary to record many tens of thousands of 'words' or pieces of information for each collision! The combination of the rate of interaction, and the quantity of information to be recorded, forces the detector to be selective in what it records. This selection also has to be performed very rapidly, otherwise spectacular interactions may be missed while a decision is still being made for a previous collision. Although the particles that are being studied are infinitesimally small, large-scale detectors are necessary to measure their detailed properties. At the collider these detectors were over 10 metres high and weighed over 1000 tonnes!

We never observe the small particles themselves but only the results of their passage through material. In this way a 'picture' of the collision and its resultant particles can be constructed. The particles we are dealing with are far too small to see, and so unless we can visualise what is happening the ideas we are discussing may seem rather abstract. Once you have seen examples of reconstructed collisions, as in figure 8.1, it is much easier to relate to this exciting world of particles that are smaller than the atom.

This chapter gives a little background on a traditional detector, the bubble chamber, which provides a clear visual record of particles

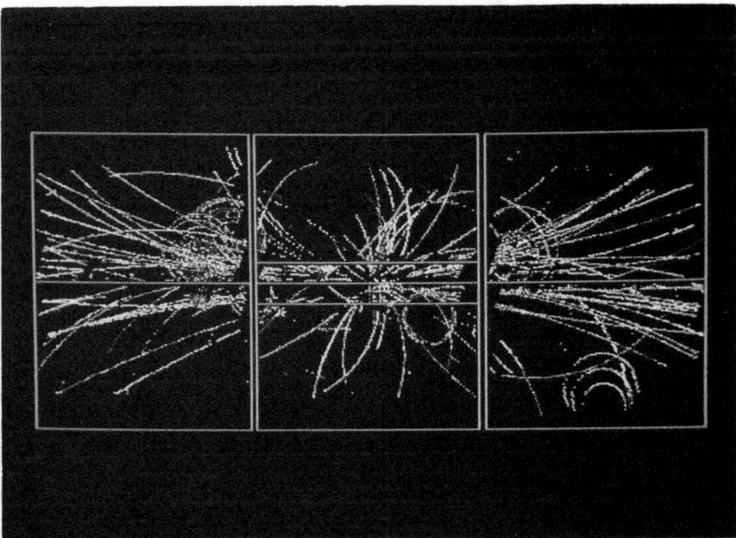

Figure 8.1. Charged particles produced by a collision of a proton and an antiproton at an energy of 540 GeV at CERN in the UA1 detector. The incident beam particles are unseen within the beam pipe but the drift chamber records the paths of charged particles produced in the collision. The particles are created at the collision point at the centre of the picture and these paths are reconstructed by the drift chamber and displayed on a monitor.

produced in an energetic collision. The weakness of this, and other similar detectors, for our current requirements is then explained. A range of electronic detectors which are more appropriate to the experimental constraints of the collider project are then introduced. These can provide even more information about the collisions than a bubble chamber. The combination of these elements to provide a comprehensive particle detector is described in chapter 9.

8.2 Bubble and streamer chambers

The bubble chamber provides the most familiar way to create a visual record of the passage of small particles. In this device, a tank containing hydrogen is cooled to such a low temperature and compressed to such a high pressure that the hydrogen becomes a liquid. In order to keep it as a liquid and prevent it from boiling to become a gas, a high pressure is applied usually by means of a piston. When a charged particle passes through the hydrogen it disturbs the atoms along its path and causes ionisation to occur. This means that the atoms are broken into positive and negative pieces, a proton and an electron. If the pressure is suddenly released, by withdrawing the piston, the liquid hydrogen starts to boil and changes into a gas. This boiling occurs preferentially where the ionisation has occurred and so bubbles of gas form along the paths of the particles. These bubbles grow in size rapidly and when they are large enough to be easily visible, a flashlight system is operated and a camera records a picture of the bubbles on photographic film. Usually several cameras are used, placed at different angles, and then the three-dimensional information about the particle paths can be reconstructed. To avoid more extensive boiling the liquid is then rapidly recompressed by the piston in readiness for the next expansion.

These bubble chambers are often several metres in diameter and as the whole device needs to be kept at temperatures of −250 degrees centigrade, the cooling and associated equipment would fill a large house. The bubble chamber has an excellent record of new discoveries and its great strength is that it can record particles travelling in all directions equally well. It also acts as the target for the collision because the incident beam particles collide with the protons of the liquid hydrogen in the bubble chamber. However, it has a serious weakness which meant it was never a candidate for use in the search for the W and Z bosons. These particles were expected to be produced only extremely rarely, in less than one in every 1000 million interactions, and the bubble chamber cannot be selective in the events that it records. As there is not enough time to record all the events, a way has to be found of selecting events that are potentially interesting before an event is recorded, which rules out the bubble chamber for this application.

A device rather similar to a bubble chamber was used for an initial survey of 540 GeV antiproton–proton collisions by the UA5 experiment. This collaboration used a streamer chamber which records a picture of a collision on photographic film rather like a bubble chamber. The streamer chamber was constructed in two parts, one on top and one below the beam pipe. Each part is filled with an ethane–argon gaseous mixture and a very high electric potential is maintained across the volume. As charged particles pass through the gas, ionisation occurs and because of the high voltage that is applied a series of small sparks (or streamers) are produced along the paths of particles. The series of short streamers is photographed with a flash system rather like the bubble chamber and recorded on film. The use of film as a recording medium for the streamer chamber has a disadvantage compared to the recording on magnetic tape used by other electronic detectors. Each picture has to be scanned and measured by a trained operator before any analysis of the collisions can be performed. As the collisions produce a large number of particles this is a difficult and time-consuming task. In the electronic drift chambers used in the UA1 and UA2 experiments, the data are recorded directly onto magnetic tape. The routine conversion of these data into particle trajectories can then be performed very rapidly by computers and only the final checks on the reconstruction need to be performed visually. None of the detectors used at the collider can make measurements inside the beam pipe and so the incident beam particles and the collision point are never recorded. The detectors record the particles as they emerge from the beam pipe and so the collision point can be reconstructed from the particle directions.

An example of a 540 GeV antiproton–proton collision as recorded by the UA5 streamer chamber is shown in figure 8.2. This experiment could not selectively record spectacular events and so had no chance of detecting the W and Z bosons. It was constructed to ensure that a visual record of the very high 540 GeV proton–antiproton collisions would be recorded, even if the two more complicated electronic detectors were not ready for the first data-taking period.

8.3 The scintillation counter

The fastest and simplest way to detect the presence of a charged particle is to use a scintillation counter. Many plastics are scintillators

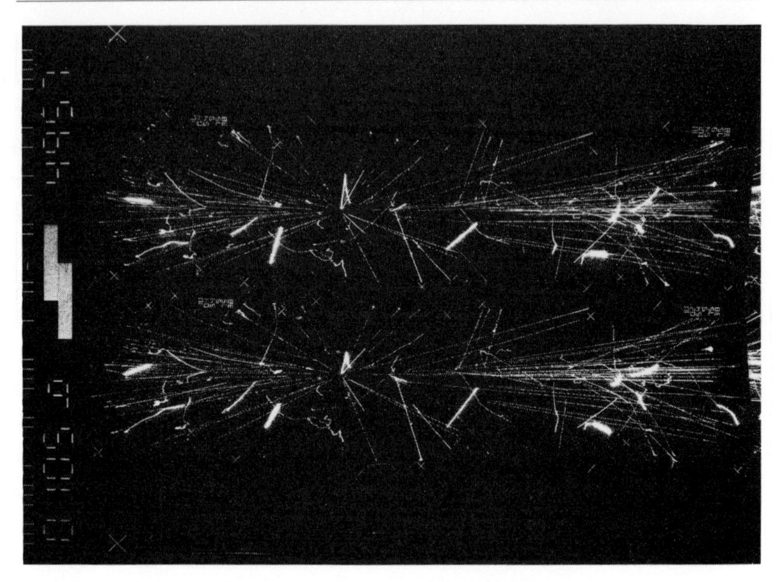

Figure 8.2. An early 540 GeV proton–antiproton collision as recorded in the large streamer chamber of the UA5 collaboration at CERN. The chamber is viewed by several cameras and these are the photographs recorded by two cameras. The crosses are marks on the sides of the streamer chamber which are used in the three-dimensional reconstruction of the events. Photograph courtesy of CERN.

and emit light after the passage of a charged particle because the atoms of the scintillator rearrange themselves and in this process visible and ultraviolet light is emitted. If the piece of scintillator is covered in black paper, then this light is reflected many times around the scintillator. This light signal is already enough to indicate that at least one charged particle has been detected but as we have to use this information very rapidly it is very important to convert this light into an electrical signal. The photomultiplier is the universal way of performing this conversion of photons of light to an electrical signal and consists of a number of metal plates held at a series of increasing voltages. When light is incident, electrons are ejected from the first metal plate and these are in turn accelerated to the nearest plate. When they collide with this plate at high speed more electrons are emitted which are in turn accelerated towards the next plate. These accelerations, over many stages, lead to a measurable number of electrons reaching the final plate and provides the required electrical signal. This can then be amplified and used in conjunction with similar signals from other scintillation counters.

One of the key features of this detector is that it can produce an electrical signal in a few nanoseconds (10^{-9} seconds) and pinpoints the passage of any detected particles to a similarly small time. This accurate measurement of the time when a particle is detected by the scintillator is extremely valuable. The exact time of the collision between antiproton–proton bunches is known and, as many of the particles produced are travelling close to the speed of light, we can predict when they will be detected at a known distance from the collision point. Most of the scintillator signals due to cosmic rays and other background processes will occur at different times and can be excluded by the time measurement. To summarise, a charged particle causes light to be emitted from a scintillator. This is collected onto a photomultiplier where it frees electrons and by a cascade process this is converted very quickly to a detectable electrical signal.

8.4 Calorimeters

Calorimeters consist of sheets of scintillator connected to photomultipliers, alternated with a metal such as iron or lead, and in this section we outline the principle features of their operation. They fulfil two important functions in a detector; the measurement of the energy of all charged and neutral particles and the identification of electrons and muons.

Most of the particles produced in an antiproton–proton interaction are hadrons, for example the proton, neutron and the most frequently produced particle, the pion. These all interact via the strong nuclear force and also the electromagnetic force if they are charged. Consider the calorimeter shown in figure 8.3, which consists of alternate strips of iron and scintillator. When any charged particle passes through the iron and scintillator it will cause ionisation to occur. In the scintillator this will lead to light being emitted which can be converted to an electrical signal by a photomultiplier. The amount of light emitted by any type of energetic particle is rather similar and so the electrical signal measures directly the number of charged particles passing through the scintillator.

If a strongly interacting hadron enters the calorimeter it will sooner or later undergo an interaction with a nucleus and produce a number of secondary particles. These will travel through the calorimeter, causing ionisation in the scintillator until each of these interacts or stops. The ionisation uses up only a small amount of energy so it is not likely to cause the primary particle to stop, but the slower secondary particles will often be stopped in this way. Notice that in all the detectors we shall discuss it is only moving particles that are detected and any incident charged or neutral hadron will generate light in the calorimeter in proportion to the number of secondary

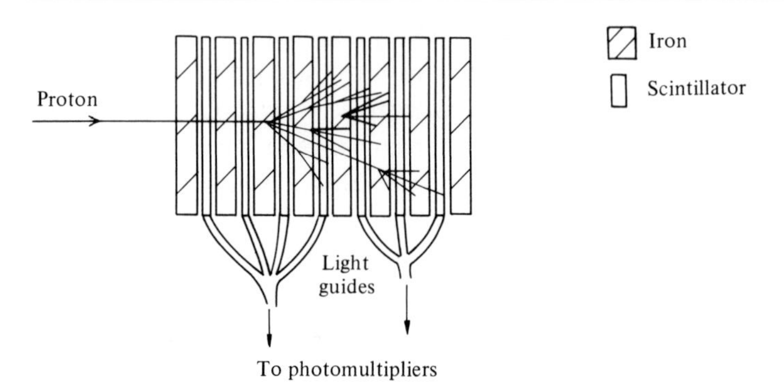

Figure 8.3. A schematic diagram of a calorimeter used for measuring the energy of a charged or neutral particle. The sheets of scintillator, which are sandwiched between sheets of iron, are connected in groups via light guides onto photomultipliers which convert light into electrical signals. An incident particle will interact in the calorimeter and produce secondary particles which will also interact. Any charged particles which travel through scintillator will generate light and the amount of collected light is proportional to the energy of the incident particle.

particles produced. This light is proportional to the initial energy of the hadron and so the calorimeter can be calibrated, by using a test beam of known energy, to measure the hadron energy. The relationship between energy and the number of secondary particles is purely statistical. High-energy particles can be measured very accurately by this method.

As well as measuring energy, calorimeters can also be used to distinguish certain types of particles. For the detection of W and Z bosons it is crucial that an electron be reliably separated from the much more frequently produced hadrons that we have just been discussing. When an electron enters material it also produces ionisation but it is much more affected by its new surroundings than the much heavier hadrons. As it moves through the material the electron experiences violent accelerations and decelerations and at each of these it emits a photon. Many of these photons will have energies above one million electronvolts. These high-energy photons lose their energy by several interesting processes, but the most popular is called pair production. In this process the photon, as it passes near a nucleus, can convert all of its energy into the creation of an electron and a positron. These newly created particles carry off any surplus energy as kinetic energy and in general do not have equal energies. Imagine the consequences in the calorimeter as an initial electron generates photons which in turn generate electron–positron pairs. These electrons and positrons also generate photons and the process develops rapidly like a chain reaction and an electromagnetic shower is produced. This cascading process happens much more rapidly than the corresponding interactions involving hadrons. This is because it is only the very light electron that generates large numbers of photons and these are not produced in hadronic interactions.

If a calorimeter is divided into segments at various depths, then we would expect electrons to deposit their energy in the early samplings and hadrons to be more likely to deposit their energy more evenly in depth. This provides the basis for the separation of the rare electrons from the much more common background of hadrons. The neutral pion (π^0) is one frequently produced hadron which could cause confusion with an electron if the calorimeter was the only detector. It decays in 10^{-15} seconds to a pair of photons and so when a π^0 is produced the photons are the objects that traverse the detector. The photons are uncharged and so do not reveal their presence in any device which detects charged particles. When the photons strike the calorimeter an electromagnetic shower, which resembles that produced by an electron, is produced. Consequently both a charged particle detector and a calorimeter are needed to distinguish between an electron and a neutral pion.

How do muons, another of our wanted particles, deposit energy in a calorimeter? Muons are charged and so cause ionisation but like the electrons they are leptons. Muons do not experience the strong nuclear force and only rarely interact in the calorimeter. As they are over 200 times heavier than an electron, they do not experience the same violent accelerations on entering material and so do not usually generate electromagnetic showers. Energetic muons pass straight through a calorimeter, depositing a small amount of energy from their ionisation in each of its layers. The penetrating nature of the muon makes the detection of muons conceptually rather simple. If we regard the calorimeter that we have just been discussing as an absorber, then the only charged particle that can penetrate it is the muon. We could put a further piece of scintillator after the calorimeter to make a measurement of the muon but it does not measure the position of a particle very accurately. At last the scintillation counter has a weakness: it produces very rapid signals, it is useful for measuring energy and is also relatively cheap but it cannot be used over large areas to measure the position of a particle very accurately.

8.5 Drift tubes and drift chambers

There are two final pieces of the detector that are required to provide a comprehensive measurement of the particles produced in

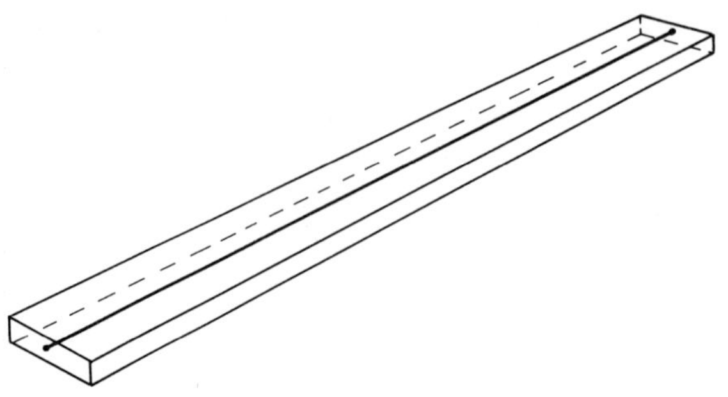

Figure 8.4. A gas-filled drift tube with a central wire held at a potential of several thousand volts above that of the aluminium case. When a charged particle passes through the tube, it ionises the gas and electrons drift to the central wire. As this drift velocity and the exact time of the collision is known, the times of arrival of the ionised electrons at the recording wires can be used to calculate exactly where the charged particle passed through the chamber.

an antiproton–proton collision. Drift tubes are used on the outside of the detector to measure accurately the position of muons, which are the only charged particles to emerge from the calorimeters. Drift chambers have the even more important task of recording the paths of all the charged particles produced in the interaction. In this section we first discuss the principles of the drift tube and then proceed to the more complex drift chamber.

We need a device that can cover an area of hundreds of square metres on the outer surface of a detector and yet still measure the position of a particle to a fraction of a millimetre. A drift tube typically consists of a tube of aluminium several metres in length with a rectangular cross-section, filled with a gaseous mixture of argon and ethane and with a fine wire along its centre (figure 8.4). The outer part of the tube is held at a negative voltage of several thousand volts compared to the wire. When a charged particle passes through the tube it causes ionisation and because of the electric field the electrons drift towards the central wire. When the electrons approach close to this wire they are rapidly accelerated, which increases the number of electrons by extra ionisation and produces an electrical signal, which is amplified and recorded. The tubes are constructed in a way that keeps the electric field uniform over most of the volume and the electrons drift with a constant velocity. The time of the electrical signal can then be used to calculate the distance of the original particle from the wire. Unfortunately it is not possible to tell on which side of the wire the track passed, but this ambiguity is usually resolved by using another set of tubes behind the first, staggered by half a tube. Another perpendicular set of drift tubes can be used to complete the measurement of the particle position.

A drift chamber is used for the final and most difficult problem, the measurement of the momenta of charged particles produced in the interaction. The simplest way to measure the momentum of a particle is by measuring the curvature of its trajectory in a region of magnetic field. We are expecting on average 30 charged particles to be produced in a single antiproton–proton interaction and so our drift chamber must be able to cope with very complicated interactions. In order to measure the momenta of the charged particles accurately it is necessary to cover a large volume around the interaction point with a drift chamber. The high-momentum particles which are of most interest are only slightly bent by a magnetic field of even several thousand gauss. Hence the large volume is essential to allow this slight curvature to be measured by recording over 100 points along the trajectory of each particle. A schematic view of a drift chamber that could be used for this task is shown in figure 8.5, where the wires are arranged in planes with each wire separated by a centimetre from its neighbour. The planes of wires are arranged radially around the interaction point and are separated by a distance of up to 20 centimetres at the outer edge of the chamber. Alternate planes of wires are held at a negative voltage of 10 000 volts and the other planes of sense wires, are held at zero potential. The chamber is filled with a gaseous mixture of argon and ethane. When a charged particle passes through the gas it causes ionisation and the electrons

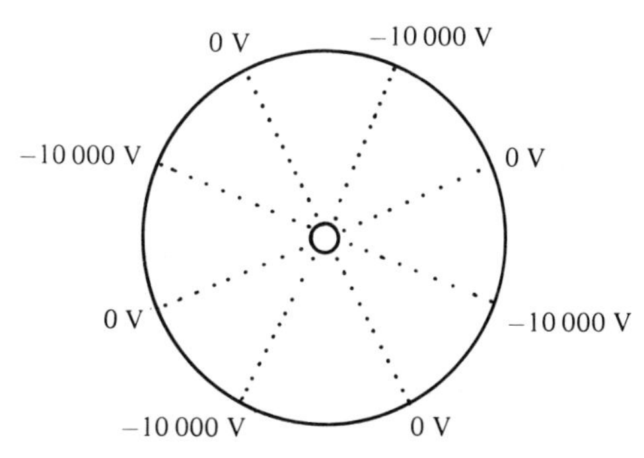

Figure 8.5. Schematic of a gas-filled drift chamber to measure charged particles produced in a head-on collision between beams travelling perpendicular to the paper. The chamber consists of planes of wires which extend for 1 metre and are held at potentials of zero and 10 000 volts. Each charged particle causes ionisation and the electrons drift towards the wires, with a potential of zero volts, and deposit an electrical signal. From these signals the paths of the charged particles can be reconstructed unless they are travelling extremely close together.

drift towards the sense wires at a uniform velocity which is independent of position. When an electron approaches close to a sense wire many more electrons are produced by a cascade process, an electrical signal is generated and the accurate time of the signal is recorded. As the exact time of the interaction is known, this allows accurate reconstruction of the position of the original track to within a fraction of a millimetre.

Several different particles may pass through this region of the drift chamber and also deposit signals on the same wires. Provided these particles are more than a few millimetres apart, the electrical signals at the wire will arrive at slightly different times and can be disentangled by sophisticated electronics. The wires are several metres in length so to measure the position of the particle along the wire, the deposited charge is measured at the ends of each wire. Equal charge measurement implies a track at the centre of the wire. After calibration, a measured charge ratio can be used to determine the position of a particle, to an accuracy of a few centimetres along the wire. In some circumstances, the total charge deposited on a wire, can be used to determine the identity of a particle. This measurement can also be used to detect whether two particles are travelling very close to one another, even if they are only recorded as a single particle in the drift chamber.

To summarise, the drift chamber records many accurate points along the trajectory of each charged particle produced in an interaction. When operated in a magnetic field, this enables the charge and the momentum of each charged particle to be measured.

8.6 Selection of interesting events

Each element of the detector produces electrical signals, which need to be recorded, in order for the results of the antiproton–proton collisions to be fully reconstructed. These signals are also required by the electronic trigger which has to decide very rapidly whether the event is to be recorded at all. The aim of the trigger is to reject events which are not caused by an antiproton–proton collision and also those genuine interactions which do not seem likely to contain a W or Z boson or other interesting particles. The W and Z boson events are very rare and so it is a very demanding requirement to reduce the common events by a large factor, without losing an appreciable number of the important events.

While the proton and antiproton beams are circulating there will be collisions between the beams and residual gas molecules inside the beam pipe, as well as those between the beams themselves. Collisions between the beams and gas molecules will generally occur well outside the detector but how can we tell that this has happened before we record the event? Scintillation counters are placed close to the beam at a distance of 10 metres on either side of the intersection region. Any charged particles hitting these counters generate an electrical signal which can be accurately timed to a few nanoseconds (10^{-9} seconds). At these very high energies all particles travel at speeds approaching that of light and so it only takes a particle a few nanoseconds to travel one metre. If an antiproton–proton collision occurs at the centre of the detector it will produce many particles travelling in various directions. Some of these will strike the scintillation counters located on either side of the detector and produce electrical signals. These signals will be generated 30 nanoseconds after the collision occurred, because each particle has to travel 10 metres to reach the counters. If there is a beam gas collision outside the detector, then some of the produced particles will enter the detector from outside. These will hit the scintillation counter on one side about 60 nanoseconds before they hit the other counter The genuine antiproton–proton interaction is characterised by a simultaneous signal in each scintillation counter whereas in beam gas collisions these signals will differ by 60 nanoseconds. The timing of the scintillation counters is used as a first-stage electronic trigger (pretrigger) to select only genuine antiproton–proton collisions for further consideration.

Now the more difficult task, how can we selectively keep the really interesting antiproton–proton collisions? There is one property of particle collisions which allows a relatively simple selection to be used in the search for W and Z bosons. Interactions between hadrons have now been studied over a wide range of energies and a very interesting property has emerged. As the collision energies are increased, the momentum carried along the beam direction by the produced particles also increases. However, the average momentum of particles in the transverse plane, which is perpendicular to the beam direction, remains roughly constant. Even when the beam energies are as large as 270 GeV, the created particles generally have a transverse momentum of less than 1 GeV! As the W and Z boson decays frequently involve transverse momenta of more than 20 GeV, then it should be possible to exclude a large fraction of the routine events without losing the W and Z bosons.

As any selection decision has to be made rapidly the calorimeter scintillators are used. Each calorimeter cell in a detector is at a known location with respect to the interaction region and so the deposition of energy in each cell records the energy flowing out from a collision into the detector. Only cells at a wide angle to the beam direction will make a large contribution to transverse energy flow unless a very large energy is deposited. The energy recorded by each cell is multiplied by a number which depends on the cell location, in order to calculate the overall transverse energy flow. The event is only recorded if the transverse energy is large. We have already seen that this calculation has to be made very quickly or else the detector will not be prepared for the next collision. As there is no time to send this transverse energy calculation to a computer it has to be performed by circuits of fast electronics so that the majority of routine antiproton–proton collisions can be suppressed. However, samples of these routine events were also recorded at intervals through the experiment because they provided a very useful calibration sample and could be used for checks of the electronic trigger.

We have already seen that the antiproton–proton bunches intersect every 8 microseconds at the detector. However, if an event is recorded, it takes over 30 milliseconds to transmit the vast amount of data onto a magnetic tape, and during this time no further event can be recorded. This means that any collision, however interesting, is missed if it occurs in the next 4000 bunch crossings! This is why it is essential to be very selective with the electronic trigger.

Whenever a selection is made, there are always two important things to remember. If the selection procedure has some logical error or involves some faulty equipment, rejected events can never be recovered. Exhaustive checking throughout the experiment was carried out to reduce the risk of this selection problem. If the trigger is too selective, then we may define requirements that are so restrictive that it will be very difficult to convince even ourselves that the observations have any significance. As the transverse energy released by a W or Z boson decay is so large compared to normal particles this is not a serious problem.

8.7 Summary

We have now discussed the main types of detectors that we used in the antiproton–proton collider experiments. There are many other detectors used in particle physics experiments but for various reasons they were not used at the collider and so these have been omitted. In the preceding review of detectors we have collected some information on how the different particles interact and how they can be distinguished.

All charged particles cause ionisation and are detected by drift chambers or bubble chambers, whereas neutral particles leave no trace in either detector. All neutral and charged hadrons, for example pions, protons and neutrons, deposit their full energy in calorimeters. With the exception of the neutral pion, these hadrons deposit energy deep into a calorimeter. In contrast an electron deposits its energy as soon as it enters a calorimeter and over a very restricted depth. A muon is the only charged particle which can penetrate a thick calorimeter and it also deposits a small amount of energy. Neutrinos leave no trace in any of the detectors that we have discussed.

An energetic electron will be identified by the following characteristics – a straight track in a drift chamber which points to a large energy deposit as it enters a calorimeter. An energetic muon will also have a straight track in a drift chamber, a nominal energy deposit in a calorimeter and a signal in the drift tubes which are behind the calorimeter.

To allow the recording of the most interesting events the background and more routine collisions have to be suppressed. This is achieved by an electronic trigger which takes the electrical signals from the whole detector and makes a rapid decision. We have now introduced the components of a particle physics detector and in the next chapter the UA1 detector is described. A description of the work required to build, test and use such a complex detector is also included, together with an outline of the steps involved in analysing the most interesting collisions.

Chapter Nine

The UA1 experiment

9.1 Design and collaboration

The UA1 experiment began when physicists and engineers who were still working on other experiments started meeting to work out a practical design for a detector to be installed at the antiproton–proton collider. Many compromises have to be made at this early stage and there is intensive activity to ensure that all aspects of the design are carefully checked. A number of prototype tests are initiated but frequently there is not enough time to digest fully all of the test results, and there are often many alternatives to be considered, without any clear advantages or disadvantages. There are always new detection techniques becoming available which can yield more accurate results than previous methods. These have to be compared carefully with well-tried techniques which will certainly work, but may be less comprehensive or accurate. The relative cost and reliability of the various designs have to be evaluated and the production schedule for each piece of the detector has to be carefully considered. In 1978 when the experimental designs were being finalised, it was hoped that the first antiproton–proton collisions could be recorded in 1981. There were three short years to convert the equipment from the drawing board to a fully tested and operational detector in the experimental area.

The collaboration forms

In 1978 the two major collaborations were still forming. The spokesman for UA1 was Carlo Rubbia who had been heavily involved in the collider project from the earliest days. He had held a joint appointment at CERN and Harvard University for a number of years and had already been the leader of many other important particle physics experiments. The UA1 collaboration gradually incorporated groups from most of the European countries and also some from the USA. These groups were from Aachen, Annecy, Birmingham, CERN, Queen Mary College (London), College de France (Paris), Riverside (California), Rome, Rutherford Appleton Laboratory, Saclay and Vienna. It also included individuals from Amsterdam, Harvard, Helsinki and Wisconsin.

This first phase of the experiment is very important indeed because if a bad choice is made at this stage, it is often very difficult, if not impossible, to correct this later when the equipment is being constructed. Members of the collaboration had worked on a number of experiments and this wide experience was particularly useful whilst the many ideas and suggestions were subjected to a critical analysis. Sometimes these designs just required common sense, in other cases very sophisticated computing and engineering checks were necessary. The end product of this design stage was a detailed proposal for the detector, which had to be submitted for approval to a committee of leading particle physicists. Most experiments to be approved use well-tested methods of particle detection or submit clear test results to show that any new ideas are really feasible. The UA1 proposal exceeded 100 pages and included a comprehensive account of the physics processes to be studied, as well as technical solutions for the various parts of the detector. Frequently there is a large difference between the optimistic expectations of an experiment and the final experimental results but this experiment was to break this rule.

In a collaborative experiment, the various parts of the detector are allocated to the contributing groups for construction and development at an early stage. Once the proposal has been accepted, these groups have the responsibility for obtaining financial support for their parts of the experiment and delivering these in working order at the appropriate time. Several groups often work together in this early phase of the experiment, and physicists from all the groups work very closely together on the experimental analysis of the data, which we will discuss in a later section.

What sort of detector should be used?

The main philosophy behind the experimental design was that a comprehensive detector should be developed, which would be able to measure particle properties over a wide range of angles and momenta. As the energy of the collider was 10 times higher than that achieved by previous accelerators, it was essential to allow for surprises, but the detector was primarily designed to measure the decay products of the W and Z bosons as accurately as possible. As the development time was very short, conventional options were chosen for many parts of the detector but a very ambitious design was used for the enormous drift chamber, at the heart of the experiment. From the beginning it was decided to produce a detector which totally enclosed the interaction region. In this way the energy of all the particles created in a collision, except those produced at a very small angle to the beams, could be measured.

In order that the charge and momenta of particles could be identified, a current-carrying coil which generated a highly uniform magnetic field was required as an integral part of the detector. There were two major options for this coil and the choice was very difficult. One idea was to have the coil shaped as a cylinder with the beam passing along its axis. Then the magnetic field acts along the axis of the coil and has little or no effect on the beam or particles travelling close to the beam. However, the chosen solution was to have a coil perpendicular to the beam, which bends some of these forward-going particles out of the beam pipe and into the detector. This is the type of choice that had to be considered for many other parts of the detector.

In such a large detector, the size of cells in the calorimeters is another important decision. If the cells are too large then there will frequently be more than one particle hitting a cell in a single collision, but if they are too small then too many photomultipliers are required and the costs become prohibitive. We have already seen that the energy deposition at various depths in a calorimeter provide important information for the identification of an electron, but how many samplings are needed and what depths should be grouped into each of these?

Many new ideas were discussed – a particularly successful development was one that substantially reduced the numbers of photomultipliers required for use in the calorimeters. A relatively new technique was adopted where a bar of plastic known as BBQ was placed in optical contact with the edge of each scintillator. When the scintillator emitted light this was absorbed by the BBQ and re-emitted equally in all directions at a different wavelength. This light could be collected from the end of the BBQ bar in a much more convenient way than from the original scintillator.

Does it all fit together?

We have now introduced a few of the ideas that were discussed while the UA1 detector was being designed. For the study of particles much too small to be seen by the human eye, a detector about the size of a normal house was being designed. However good the original idea, it will all count for nothing if the final detector does not match its specification, so now the ideas had to be translated into a real detector weighing over 2000 tonnes. There were also important constraints on the time, money and manpower that could be used to construct the detector and get it working correctly. The work was shared out between the collaborating groups in a way that matched as well as possible the available expertise, financial support and manpower. The civil engineering to create the new underground areas and beam transfer tunnels was the responsibility of the CERN laboratory.

All of this work had to be carefully coordinated so that there were no missing items and everyone knew clearly when everything was expected to be ready. Many of the detector elements were critically dependent on the delivery times of manufacturers of scintillators, advanced electronics and numerous other items. Carlo Rubbia was involved in all the major planning decisions but two coordinators handled many of the day to day problems and organisation. One of these was responsible for the UA1 budget and the other dealt with the internal communication between the participating groups and also coordinated the installation work in the experimental area. Although the project was handled separately by the various groups, each of them needed to be aware of all the repercussions of any modification to the original design. There was only a limited space available in the experimental area and so it was important to optimise the arrangement of detectors to achieve the highest detection efficiency.

9.2 Construction of the detector

Once the experiment had been approved the various groups started to prepare their pieces of the apparatus. Before the final items could be order or constructed, each part of the apparatus was studied in even more detail as rapidly as possible so that orders could be placed in good time. There were regular collaboration meetings to ensure that any useful information was distributed efficiently. When the various designs were complete, each group made a more detailed proposal for their part of the detector and submitted this for approval in their own country before the construction could be started.

Central detector

One of the most important and technically advanced items was the large drift chamber, called the central detector. The objective of this chamber was to provide the information to reconstruct the directions and momenta of all charged particles produced in the antiproton–proton collision. In order to do this efficiently, a huge drift chamber with a volume of 25 cubic metres was constructed. This cylindrical detector, 6 metres in length and over 2 metres in diameter, was the responsibility of the CERN group in the experiment. The detector is filled with a gaseous mixture of argon and ethane and contains over 12000 wires. Half of these are held at zero voltage (sense wires) and the other half are held at a negative voltage of 30000 volts. Any moving charged particle causes ionisation of the gas and the resulting free electrons drift towards the sense wires with a uniform velocity. The electrical signals are then amplified and analysed to measure the arrival time and the deposited charge of each passing charged particle. The wires in the central detector are all arranged in the same direction which is perpendicular to the beam particles. The sense wires are mounted at one-centimetre intervals in rows which are separated by 20 centimetres from similar rows of high-voltage wires. Alternate rows of these wires are arranged throughout the volume of this drift chamber. These planes are arranged differently in the various parts of the central detector (figure 9.1) to maximise the amount of information recorded about each particle.

There were several very difficult problems to be solved as the wires in the drift chamber have to be kept under tension, as on a guitar, which requires the outer shell of the chamber to be mechanically strong. However, all the particles produced in an antiproton–proton collision need to traverse the shell of the central detector before any energy measurement can be made by a calorimeter. For this reason it is important to make the outer shell very thin and so a compromise has to be made between its strength and weight. The French firm Aerospatiale, which specialises in producing devices for space research, won the contract for the production of the shell of this detector. A light shell was finally used, which was prestressed so that it held the wires under tension automatically because its walls naturally flexed outwards and kept the wires at the correct tension. The volume of the chamber was divided into six separate half-cylinder-shaped

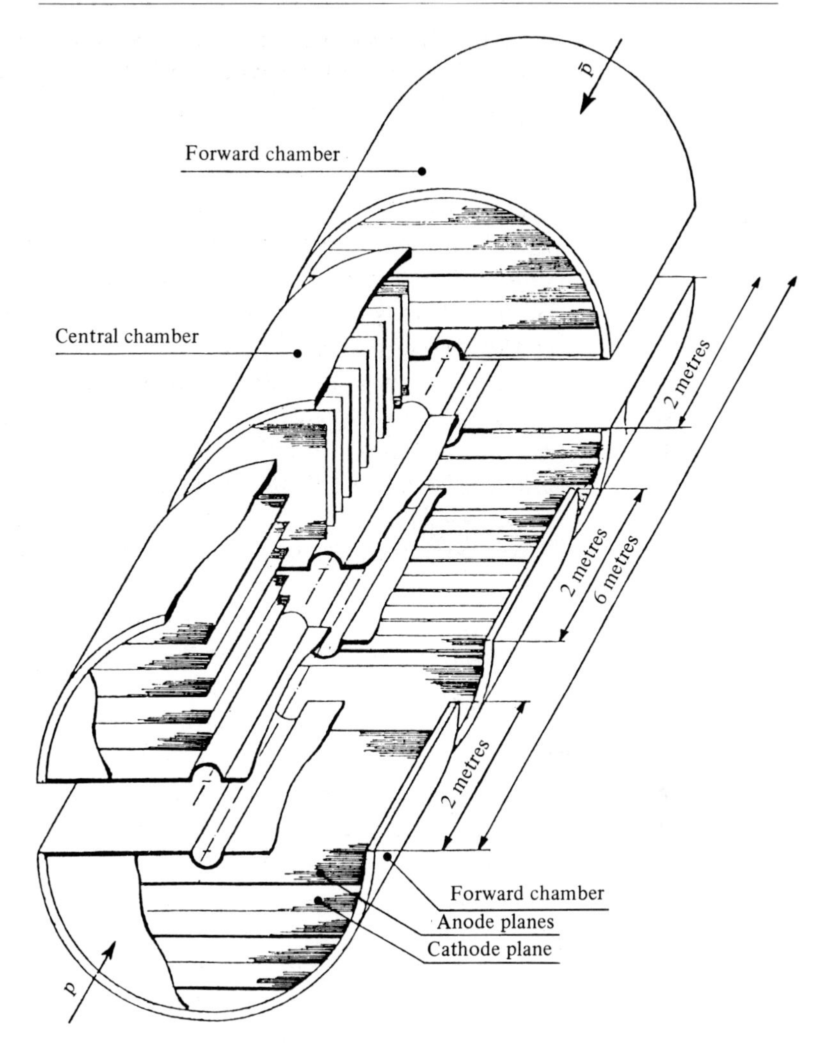

Figure 9.1. This gas-filled central detector used in the UA1 detector consists of six main modules containing over 6000 sense wires for detecting electrical signals. The wires are all arranged in planes separated by 20 centimetres and all wires are parallel to the magnetic field. The electrons produced by the ionisation of charged particles drift to the low voltage sense wires and the passage of these particles can be reconstructed from the electrical signals.

modules and as this was a key element in the whole detector, two spare modules were also constructed.

The high-voltage wires, which needed to be held at 30000 volts, presented their own problems as it is very difficult to maintain such a high voltage without very great care in design. The electronics required to read out the signals posed another challenge, as there

Figure 9.2. Final electronic units being added to one half of the central detector. It is important to mount some of the electronics on the chamber itself to avoid losing accuracy by transmitting along long cables. Photograph courtesy of CERN.

were 6000 sense wires in the chamber and signals need to be amplified and processed from each of these. There was very little space available on the outside of the central detector but most of the electronics had to be placed there and the final track detector was totally enclosed by electronics (figure 9.2).

Electromagnetic calorimeter

The central electromagnetic calorimeters are used for the identification and accurate energy measurement of electrons and these were the responsibility of the groups from Saclay and Annecy. It was decided to use a calorimeter of alternating layers of scintillator and lead, each 1 millimetre thick, and as the requirements were for a very accurate device it was a formidable challenge. Light is emitted by the scintillator when any charged particle traverses the calorimeter and this is absorbed and re-emitted by BBQ bars. Plastic light guides attached to the end of each BBQ bar collect the light from many separate scintillator sheets onto a single photomultiplier where it is converted into an electrical signal (figure 9.3). Different cell shapes were used for the calorimeter around the sides (gondolas) and ends (bouchons) of the central detector but in each case the signals

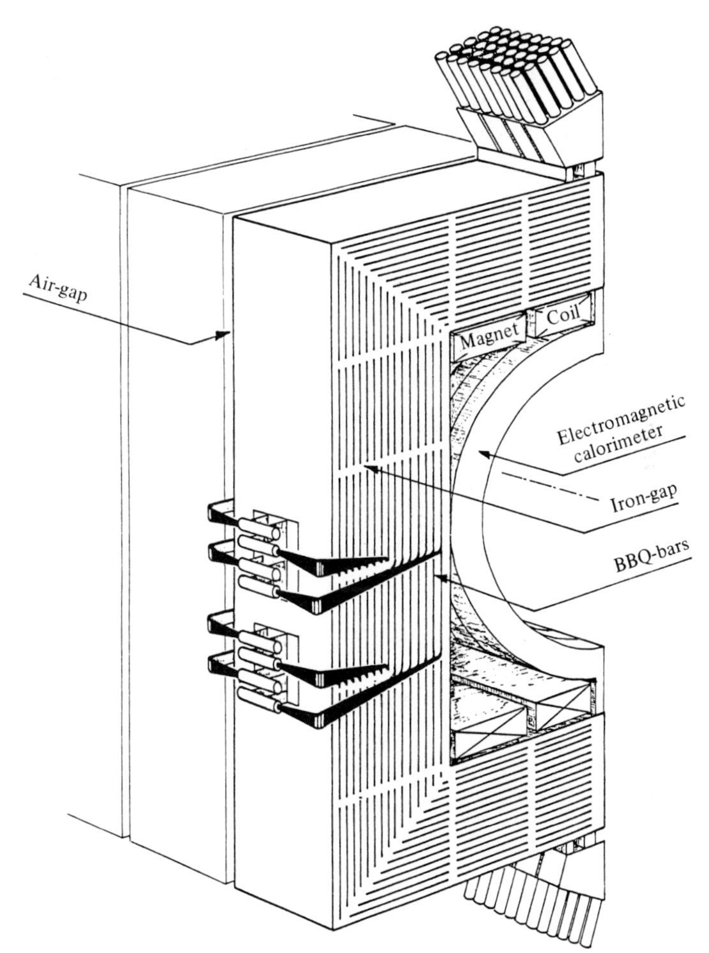

Figure 9.3. A schematic view of part of the central calorimeters which surround the central detector on both sides. The inner electromagnetic calorimeter (gondolas) are shaped to fit closely around the central detector but the hadron calorimeter (Cs) are rectangular to leave room for the coil of the magnet. The hadron calorimeter signals are divided into two depth segments as shown whereas the electromagnetic calorimeter is divided into four. In each case the light deposited by particles is detected by sheets of scintillator, absorbed and re-emitted at a different wavelength by BBQ bars and transmitted via light guides to photomultipliers.

from the photomultipliers could be used to work out the energy and position of the energy deposited in each cell. In order to aid the identification of electrons, the energy deposition was recorded at four different depths (samplings) in this calorimeter. The design of the electronics for the electromagnetic calorimeters was the responsibility of the Vienna group. There is a wide range of energies to be measured, as the energy deposited by a muon is only a few hundred MeV, whereas an energetic electron may deposit up to 100 GeV in a cell. The electronics must be able to make a precise measurement over this large energy range.

The electromagnetic calorimeters were extensively studied in test beams and further energy calibrations were performed with a very intense 7-Curie Cobalt 60 radioactive source! This was also used to monitor the variation in energy response over the surface of these detectors. When this intense source is used the levels of radioactivity are very high and so these scans have to be fully automated. The performance of the thousands of photomultipliers are regularly monitored by using a laser calibration system. Light can be transmitted from a laser to each photomultiplier via a very thin light fibre. This known signal is used to monitor the stability of each photomultiplier and the laser stability is in turn monitored by a weak radioactive source, which decreases in activity at a precise rate.

Magnet

A uniform magnetic field of 7000 gauss is generated over the volume of the central detector by passing 10 000 amperes through the thin aluminium coils of a 800-tonne electromagnet. The coils are situated outside the electromagnetic calorimeters, providing a horizontal magnetic field perpendicular to the beam direction, with the magnetic flux returning through an iron calorimeter which surrounds the coil. With this arrangement, a very uniform magnetic field is produced in the volume of the central detector and the magnetic field outside the detector is kept very small, which enables sensitive photomultipliers to be operated successfully. The magnet was designed and constructed at CERN and was fully assembled to allow detailed measurements of the magnetic field well before the underground area was ready. These measurements confirmed that an extremely uniform magnetic field had been successfully produced. A side view of the assembled UA1 detector is shown on figure 9.4.

Hadron calorimeter

The outer iron of the magnet played a dual role as it was also used as a hadron calorimeter. Energetic hadrons can penetrate much deeper into material than electrons or photons and deposit most of their energy in this hadron calorimeter which surrounds the electromagnetic calorimeter. The iron of the magnet was divided into C-shaped modules (figure 9.3) surrounding the gondolas and rectangular I modules enclosing the bouchons. The instrumentation of this calorimeter was the responsibility of the Birmingham, Queen

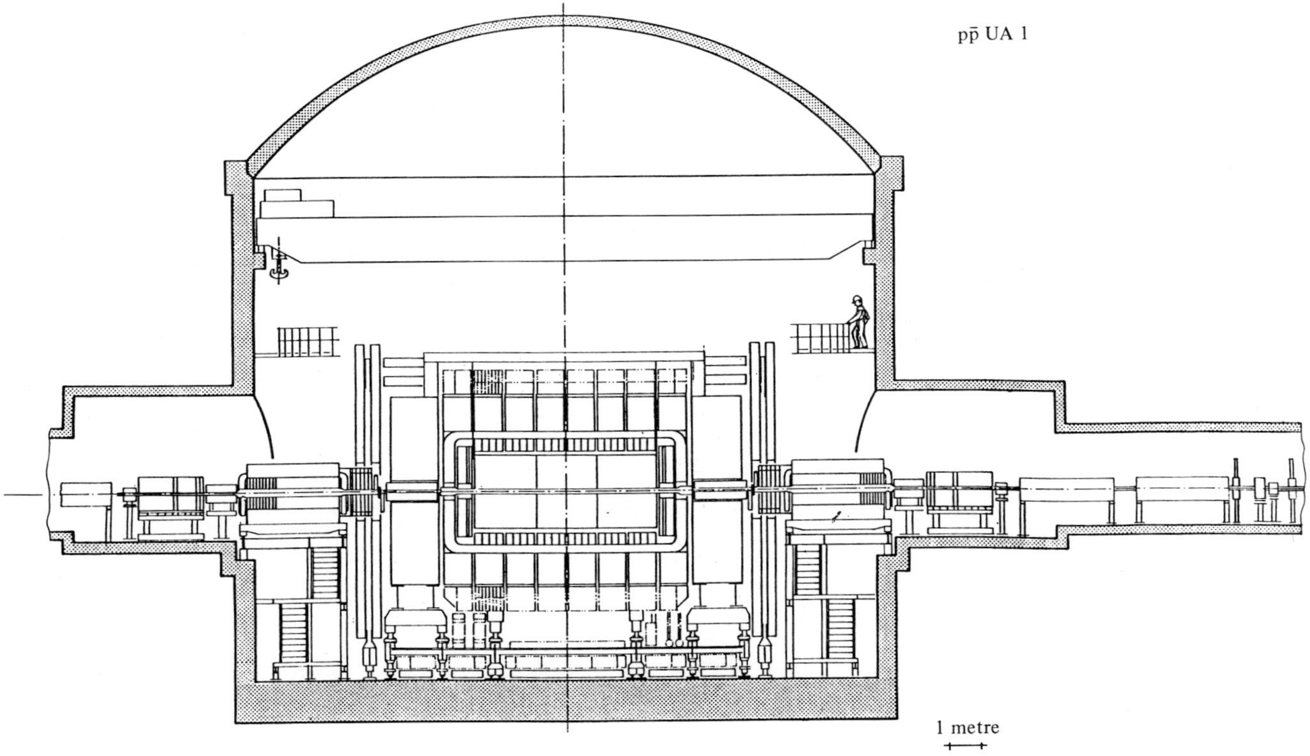

$p\bar{p}$ UA 1

1 metre

Figure 9.4. A side view of the entire UA1 detector when installed in the super proton synchrotron (SPS) tunnel. The incident beams travel along the central beam pipe and collide at the centre of the detector. In each direction particles traverse the central detector, followed by electromagnetic and hadronic calorimeters and finally the muon chambers. Particles produced at small angles are detected by the forward and very forward detectors. The whole 2000-tonne detector is mounted on a rolling chariot so that it can be withdrawn from the tunnel for normal SPS operation.

Mary College (London) and Rutherford Appleton Laboratory groups. Each element was made up of alternating sheets of 5 centimetres of iron and 1 centimetre of scintillator and the light from the scintillator was again transmitted to photomultipliers using bars of BBQ placed along each edge of the scintillator. The hadron calorimeter was subdivided into several hundred separate cells, each covering an area of half a square metre, with the energy deposition recorded as two separate depth samplings.

Careful studies were made of the performance of the various elements of this calorimeter in order to optimise the amount of light received by the photomultiplier for a given deposited energy. A very useful calibration tool for all parts of the apparatus, including the calorimeters, are the cosmic ray particles which are continuously bombarding the earth. These include muons which deposit energy in a calorimeter in a uniform and predictable manner and provide a continuous calibration of the detector. The mounting of scintillator, BBQ and photomultipliers to the iron of the Cs and Is was performed in a building close to the proton synchrotron (PS). When this is running, additional high-energy muons are produced which pass through the concrete walls surrounding the accelerator. This flux is very low and does not constitute a significant increase over the cosmic rays but these additional muons were also used in the early calibration of the hadron calorimeter. At construction a thin light fibre was connected to each sheet of scintillator and these fibres were connected to a nitrogen laser which emits ultraviolet light. This laser system was used to monitor any variations in the calorimeter performance and the laser stability was again monitored by radioactive sources.

While the calorimeters were being constructed, a large support structure was built to enable elements of each calorimeter to be studied in test beams of particles from the PS and the super proton synchrotron (SPS). Each element of these detectors weighed several tonnes and some were not self-supporting so a carefully engineered platform was required. The aim was to arrange for each of the calorimeters to receive beams at known energies so that the signals produced in the calorimeter cells could be fully investigated. The electromagnetic and hadronic calorimeters were tested individually and also in combination as they would be used in the experiment.

Separate data were recorded for incident muons, electrons and hadrons at various energies and angles. These tests were carried out over many months and more than one million interactions were recorded on magnetic tapes. A preliminary analysis of these data was performed using the small data acquisition computer, but the detailed analysis was performed using the large computers at CERN and other laboratories. The experimental data were then compared to the predictions of 'Monte Carlo' programmes, which try to reproduce the physical processes occurring in the calorimeter with a computer simulation. Once these programmes have been tested and verified with experimental data they can be used to predict the behaviour in awkward corners of the detector where test measurements are not really feasible. In this way the total calorimeter can be calibrated so that in the experiment it can be used to measure energies reliably. These round the clock beam tests were often conducted in very cramped conditions, with several teams operating in parallel with limited computer facilities. However, these periods were very productive and indeed crucial to the accurate measurement of the energy in the UA1 detector.

Forward detectors

We have now discussed the drift chamber and electromagnetic and hadronic calorimeters which occupy the central region of the UA1 detector. However, particles that are produced at angles of less than five degrees to the beam are not 'seen' by these detectors. A set of forward calorimeters and drift chambers using similar techniques to those described extends this coverage so that the energy flow is measured down to a fraction of a degree. These were the responsibility of the College de France, Rome and Riverside groups. Many of the particles created in high-energy collisions are produced at small angles to the beam, which makes detailed measurements very difficult as many particles are travelling close to each other. At a distance of 25 metres from the intersection point, a small set of drift chambers and scintillators was installed, which could be moved to within a few millimetres of the beam. This was used for the measurement of scattering at very small angles and also as a monitor of the beam intensity.

Muon detector

The muon detector consists of drift tubes which provide the outer layer of the UA1 detector and were designed and tested by the Aachen group. The muon chambers were tested in Germany and were then transported by rail to Geneva and so the maximum size of the individual chambers was limited by the size of the railway tunnels! Over 50 modules of muon drift tubes were constructed which required 30 kilometres of extruded aluminium! Every 4 metre by 6 metre module has four crossed planes of drift tubes, each with a

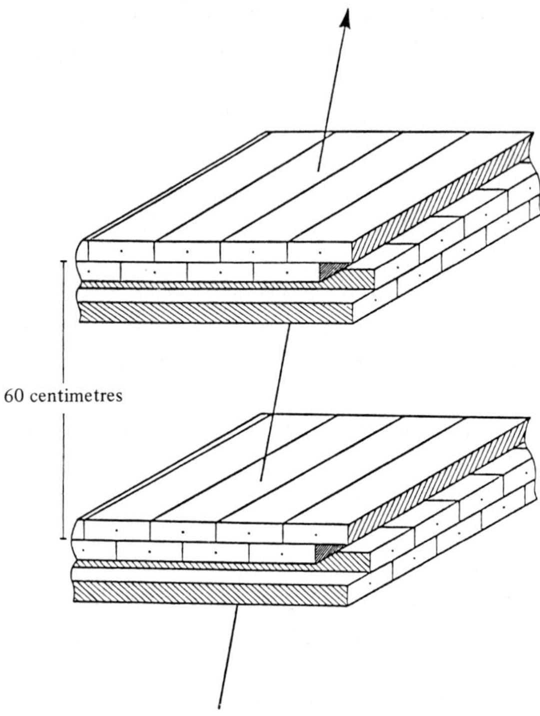

Figure 9.5. The muon drift tubes are connected together to form a muon module which consists of two separated units. Each unit comprises two pairs of crossed drift tubes with alternate layers staggered by half a tube to remove ambiguity about the track location. With this arrangement it is possible to measure the muon track position and angle very accurately after it leaves the outer calorimeter.

cross-section of 5 by 15 centimetres (figure 9.5). A sense wire located along the centre of each tube detects the ionisation caused by the passage of a charged particle and the tubes are filled with the same gas mixture as the central detector. The information recorded by a muon module allows a precise measurement of the position and direction of the muon track after it has traversed the calorimeter.

Electronic trigger

A selective muon trigger for the experiment was constructed by the Aachen group. Genuine high-momentum muon tracks from an antiproton–proton collision generate signals in muon chambers which point approximately to the interaction point. Electronic processors were designed and constructed to process the muon signals and to decide in less than one microsecond (10^{-6} seconds) whether these hits were consistent with an energetic muon. The number of background particles was further reduced by requiring that a hadron calorimeter cell was also active in the nearby zone of the detector. This muon trigger was used throughout the running periods and was responsible for about half of the data recorded on the magnetic tape. An even more selective muon trigger was developed by the group from Amsterdam, using several microcomputers working in parallel.

The final piece of electronic hardware for the experiment was that needed to select the most interesting calorimeter triggers, including the W and Z boson decays to electrons. This was designed by the Birmingham, Queen Mary College (London) and Rutherford Appleton Laboratory groups and was constructed and tested at the latter laboratory. This trigger processor was designed to enable a decision to be taken in a few microseconds about the desirability of each event. As described earlier the signals from calorimeter cells could be converted to a transverse energy for all cells or localised groups of cells. Only one-thousandth of the collisions could be recorded so the trigger processors played a crucial role in the data taking.

Once the various elements of the detector had been constructed and tested, the assembly was started. As the full detector weighs over 2000 tonnes and had to be assembled 30 metres underground, this formidable operation was a real challenge. If there were any misunderstandings or elements with the wrong size, it would soon be very obvious. Every piece of the detector had to be transported to the intersection region and then lowered by crane to the underground area to build up the complete detector shown in figure 9.4.

9.3 What computers are used?

We have discussed in some detail the accelerators and the experimental detectors that were needed in the search for the W and Z bosons. There would of course be no results from the experiments at all without the dedicated efforts of the physicists, engineers and technicians at all stages of the project. It also needed extremely powerful computers to record, digest and check the immense amount of data recorded in the collider experiments. A large fraction of the work performed by the physicists in these experiments involves the use of various types of computers. In order to explain the day to day running of the experiments, we must first describe the computers that were used in this project. Once this has been done it will be much easier to follow the description of the search for W and Z bosons.

The aim is not to give technical details of the computing but rather the flavour of the types of tasks that are crucial to the success of an experiment.

Data acquisition

The most important computer for any experiment is that used for data acquisition because it is responsible for transmitting the information recorded by the experimental detector onto a magnetic tape. Any problems with this computer are extremely serious, as they prevent the experiment from accumulating new data. Complex detectors record information from tens of thousands of wires and photomultipliers and for each selected antiproton–proton collision (event); the electrical signals from all of these elements have to be recorded accurately onto magnetic tapes. The information for a single event occupies typically 50000 'words' of computer memory, about the same number as in a telephone directory. Under normal beam conditions, 1 kilometre of high-density magnetic tape was filled with data every 10 minutes in the UA1 experiment and each tape contained the recorded information from 1000 carefully selected collisions.

This same computer is used to load instructions into the electronic triggers used in the experiment. Whenever a calibration of the many parts of the apparatus is in progress, it is controlled by the same computer and the results are also recorded onto magnetic tape. As there are so many elements to check in such an experiment, there are many periods when the computer is very busy indeed. This is especially true in the frantic days just before the data-taking periods are due to start. During data taking, this computer is also used to monitor the progress of the experimental run by performing checks on the recorded data and verifying that all parts of the detector are operational. All of the programmes which are used for these tasks are written and tested by the physicists on the experiment, with the help of a few computer specialists. The data acquisition computer, although very powerful, cannot cope with all the detailed checking of the apparatus that is really necessary.

This task of checking the data during the experimental run cannot be overstressed. The available time in an accelerator experiment is very short and so any losses caused by problems with the apparatus or computers cannot usually be recovered. As there are so many elements to be checked it is often possible for some problems to go undetected and physicists have the responsibility during the run to use every available method to detect errors or inconsistencies in the data. The electronic information is passed from the detector along cables to the data acquisition computer and on the way it can be interrogated by small microcomputers. In this way some of the simple checking of the various parts of the detector can be done by small computers and the main computer can be left to its primary task of recording the data.

Access to the data acquisition computer and the microcomputers is mainly via terminals in the experimental control room. This is a room at ground level, above the underground experimental area, where the detector is recording the experimental data. There are up to 20 computer terminals together with a number of video display screens where the recorded information can be displayed, in this 10 by 3 metre room, which is the key control room for the experiment during running time. At critical times this room was filled to capacity, but more typically during data taking, the room would contain eight people, each responsible for a specific part of the detector. Many of the terminals are similar to those used on home computers but others, known as touch screens, allow the choice of options by a single finger touch, which is much faster than using typed commands.

We saw earlier that an electronic trigger was essential to reduce the rate of data accumulation in the experiment, otherwise there was a risk that the data acquisition would be unavailable when the very rare W or Z bosons were produced. However, this trigger decision is final, and once an event is rejected it is lost forever, and so experimenters are very cautious with any selection criteria. In order to obtain early results we are interested in fully analysing the recorded events very rapidly, and the main steps in event analysis are outlined in the next section. However, there is not enough computer power at CERN and the collaborating laboratories to achieve this fast analysis of each recorded event. This can be illustrated by noting that events are being recorded around the clock, once every second. However, it takes over 20 seconds on the most powerful computer to analyse fully each event, and so to achieve rapid results a further event selection has to be made.

In the UA1 experiment this was done by passing the data through four parallel computer processors before the events were written onto magnetic tape. These processors are called emulators and are very powerful computers in their own right but are much cheaper than standard computers because they only contain the processing part of a computer without the usual input and output devices. The emulators worked very well and were a crucial element in the rapid analysis of the UA1 experimental data. These processors applied even stricter cuts to the data than the electronic triggers and diverted the most interesting candidates, about one-tenth of the data, to special tapes. These were called 'express line' tapes and were fully analysed within a few days of being recorded. A copy of these special events, and those not chosen by these emulators, was recorded on the standard date tapes of the experiment and most of these events were processed in the months following the data-taking period.

The more sophisticated checks of the data cannot be implemented on the data acquisition computer because it is too busy but at the same location there is another powerful computer which is used to perform some of these other checks. This computer is used to produce distributions of the wires or the calorimeter cells that are active and also more detailed correlations of the results from different parts of the detector. If an error is detected with these programmes then it probably means that data taken during the previous hour has been affected but without this rapid check, a much more serious loss of time could be incurred.

All of the computers that we have discussed are linked together by a communications network, which means that information can be exchanged between them and that a single terminal can be used to communicate with several different computers by dialling a numerical code on a switch at the side of each terminal. The computers that we have mentioned so far are dedicated to the UA1 experiment, but at CERN there exists a very powerful central computing facility which is shared by all the experiments carried out at the laboratory. This is housed in its own large building in the middle of the CERN site and is one of the largest computer installations in the world. From the visitors' gallery there is a good view of this whole complex, which operates round the clock, seven days a week with over 30 separate tape units, enormous magnetic disk capacity and two independent computer processors. Below this main computer room, tens of thousands of tapes are stored in a large tape library and are despatched to the computer room via a conveyor belt when needed for a computer program. These major computers are the only ones powerful enough to achieve a rapid analysis of these complicated experiments, and during the collider runs the UA1 and UA2 experiments made full use of the priority they were given on these computers. It would have taken much longer to extract the important physics results from these very complex detectors without these central computing facilities.

In the same building as the central computers there is another very large area which is used to maintain contact with the outside laboratories. Most of the universities and research laboratories involved in particle physics experiments also use their own computers.

In most cases these are connected to worldwide networks, which enable results and programs to be exchanged between the collaborating laboratories, but in general these links are not fast enough to allow large quantities of data to be transmitted in a reasonable time. Most of the experimental data are still transferred by carrying magnetic tapes back to the home laboratory, although some successful tests of transmission of data via satellite have been performed. These outside computers are very important because they allow analysis and programme development to be done by people at their own institutes. This computer power can also be used to supplement the CERN-based computers and allow more rapid analysis of some classes of events.

The UA1 collaboration was made up of 150 physicists from many different countries. The use of networked computers enabled this large number of people to keep in touch with each other and up to date with the latest news from the experiment. Any important results or announcements could rapidly be disseminated to the whole collaboration and without this electronic mail the organisational problems would have been immense. The computer programs were also well organised and documented so that many parts of the analysis were developed separately and finally merged together to produce a complete set of analysis programs.

9.4 Raw data become reconstructed collisions

We have mentioned several times the full analysis of an event and pointed out that this procedure takes much longer than the recording of the data. What is involved in this analysis and why does it take so long? In this section we will outline the most important steps without worrying about any of the technicalities. The raw information that is written on tape for each event records the activity in each element of the detector at the time of that collision. This will include the electrical signals recorded on each of the thousands of photomultipliers and drift chamber wires. The analysis program converts this huge array of uncalibrated numbers into a reconstruction of the particles produced in the energetic collision. Once this is achieved we can begin to search for signs of the electrons and muons which could result from the production of W or Z bosons.

The first stage of analysis is to calibrate the raw signals from each cell so that they represent a known amount of energy. This requires the results from test beam studies and the results from the various laser calibrations which are performed regularly. Once this stage is done, the energy in GeV deposited in each sampling of each calorimeter cell is known, but what about the signals from the drift chamber wires? There is so much calibration to do in this case that some of it is done by microprocessors before the data reaches tape, and these calibrations are continuously monitored to remove any effects caused by small changes in the gas composition in the central detector. The most accurate information recorded by the wire is the time of arrival of the signal caused by the passage of a charged particle, which needs to be converted into a distance of the particle from the wire. This requires several extra pieces of information: the exact 'zero time' of the collision and the velocity and drift angles of the electrons in each of the chambers. Using calibration constants the accurate particle position is calculated, and finally the amount of charge recorded at each end of the wire is converted into a measurement along the wire.

Once the raw signals have been calibrated, they are used to reconstruct the trajectories of all the charged particles produced in the collision. The thousands of hits in the central detector are sorted so that signals which could result from a single charged particle or track are grouped together. Where a charged particle is produced in an empty region of the chamber this is a straightforward procedure, but there are often over 100 charged particles produced. Many of these travel in close groups or jets of particles and then it is often very difficult to allocate unambiguously a hit to one of several close tracks. In the next stage of analysis, the computer performs a track fit to the points which have already been separated into track candidates. A curve is superimposed on the points and its curvature and direction are altered until the line follows the points in the most accurate way. As the magnetic field is known, the curvature of a track reveals the momentum and the electric charge of this particle. Most of the reconstructed tracks will point back to a single vertex, which identifies the antiproton–proton collision point. However, not all of the tracks originate from this vertex because they can result from the decay of an unstable particle or from the interaction of one of the particles produced in the original collision.

We have now reconstructed the collision point, the electric charge and momentum of each charged particle produced in the collision and the energy deposited in each calorimeter cell. The calorimeter cells record the energy of both neutral and charged particles, and so in principle the energy of the neutral particles can be deduced. This works well if there are no charged particles pointing to a cell as then all the energy has to be due to neutral particles. However, if charged and neutral particles both strike the same cell, then it is usually very difficult to separate the energies reliably and it is particularly hard for low-energy particles, where the fluctuations in the energy deposited in a calorimeter become very large.

The hits recorded in the outer muon chambers, behind the calorimeter, are also calibrated and combined to produce candidate tracks in these chambers. As the muon chambers are outside the region of magnetic field these are straight tracks with a direction but producing no direct measurement of particle momentum. Before the information from these different detectors can be combined, it is essential that the relative positions of the individual elements are accurately known. These geometrical measurements are obtained from careful surveys of the detector and cross-checked by the use of cosmic ray muons. These particles are very suitable for this task as they are continuously available and they deposit energy in the outer drift tubes, calorimeters and central detector.

Selection of candidate events

We will now briefly outline the steps involved in extracting energetic electron and muon candidates from these reconstructed events. An energetic electron will have produced a very straight track in the central detector and we can predict where this would enter the electromagnetic calorimeter. If it is an electron then it should deposit all of its energy in a localised region near its entry to the calorimeter. For an electron this amount of energy should agree with the momentum of the particle as measured in the track chamber and we can make further detailed checks on the energy deposited in the calorimeter. This energy should be deposited exactly where the particle hits the cell and there should be no energy penetrating to the hadron calorimeter or even the deepest part of the electromagnetic calorimeter. If a particle satisfies these and further technical requirements then we can reliably identify it as an electron.

How do we recognise the energetic muon? The requirement on the track in the central detector is the same as for the electron, i.e. it should be very straight. If it is a muon the particle will deposit a limited amount of energy in the calorimeter cells that it traverses, before entering the outer muon drift tubes, which are more than 6 metres away from the collision point. There is a magnetic field in the central region and also a similar field in the iron of the hadron calorimeter. In order to predict the arrival point of the muon at the outer chamber the effect of these magnetic fields on the particle trajectory have to be calculated, together with the uncertainties in this extrapolation. Once this is available the extrapolated track values can be compared with those measured in the muon chambers and if the agreement is good, and there is evidence for calorimeter energy along the track, then this particle becomes a muon candidate. Further stringent checks on the quality of the central detector track are required before it is verified as a muon.

We saw earlier that the detector completely surrounds the interaction region down to a fraction of a degree from the beam direction. In each collision we expect the momentum to be conserved along the beam and in the two perpendicular directions in the transverse plane. As some particles escape detection by passing along the beam pipe, it is impossible to check the momentum balance along the beam direction. However, in the transverse plane the momentum balance is relatively unaffected by particles escaping down the beam pipe as these cannot carry large transverse momenta. So transverse momentum balance can be used in a search for particles which escape from the detector without interacting. Each calorimeter cell energy deposit and its known position are combined to calculate an energy flow in the transverse plane. When this is repeated for all the cells the missing transverse energy for the event can be measured. There are always fluctuations in the energy deposited in calorimeter cells but typically this leads to an imbalance of only a few GeV in the transverse plane. When the missing transverse energy is much larger than this, then it can indicate that some particle or particles have escaped detection. The weakly interacting neutrino becomes a strong candidate as the carrier of this missing transverse energy.

Before we leave the analysis we should outline the procedure for extracting jets of charged particles or calorimeter energy from the detected information. A similar technique is used in both cases, which collects the tracks or cells which are nearby in space into a single jet. The energy and momentum of this jet can be calculated by combining the contributions of the individual tracks and cells.

We have now done our best to reconstruct the particles that are produced in an antiproton–proton collision. The energetic electrons, muons and neutrinos have been identified and the energy flow has been localised into jets but are there any further steps before the results can be published? This is really just the beginning of the event analysis because these routine procedures deal with the majority of the repetitive calculations, but the selected events have to satisfy much more stringent checks before they are used in a scientific publication. Some of these checks are just extensions of the earlier discussion and are performed by running further computer programs. However, there is one distinctly different check that is always made, and this is a study of the 'picture' of the collision. As this formed an important element in the analysis of all interesting events, it will be described in some detail in the next section.

9.5 'Picture' of a collision

Finally the computers used to create high-precision displays of the recorded events have proved invaluable in these experiments. After all the automatic checks and computer reconstruction it is very reassuring that a physicist can still spot inconsistencies and trends that have not been detected by more automatic methods. We have already described how the information recorded about an event can be used to create a computer-reconstructed version of the proton–antiproton collision. On a visual display the raw signals recorded by each wire and photomultiplier can be displayed at their known locations together with the fully reconstructed collision. A complete visual picture of the collision can be produced on a high-precision display, containing as much or as little information as required. This is a major advantage over a bubble chamber photograph where it is not possible to simplify the contents of the picture. As the analysis process we have described yields reconstructed tracks directly in all directions, this means that the 'picture' of the collision can be inspected from any chosen angle in order to get the best view of the collision.

Once the most promising events have been fully processed, they are then rapidly transferred to the display computer. At CERN there are one colour screen and two black and white screens dedicated to the analysis of the UA1 experiment. The monochrome screens can display points at 16 million different locations, which allows a very high-precision inspection of any part of the detector. Although the colour screen has a slightly lower resolution, this is more than compensated for by the presence of colour. Once the information is loaded into the graphics terminal, the picture can be manipulated rapidly in a variety of ways. The device includes microprocessors which enable magnification, movement and rotation of the picture to be performed very rapidly under the control of the operator, which makes it easy to inspect rapidly many different properties of an event.

Among the many options there are several that have proved invaluable in the verification of events, including the automatic selection of tracks with large transverse momenta. Most particles are produced along the beam directions and it is relatively rare for a particle to be detected with high transverse momentum. On first inspection the events recorded at the collider, with as many as 50 charged particles created, seem very complicated. However, once the particles are required to have more than 1 GeV of transverse momentum, then in the vast majority of events very few tracks are displayed and as the tracks of most interest have transverse momentum above 20 GeV, they are very striking. Colour can also be used to highlight the most energetic tracks and calorimeter cells, and hence is very helpful when trying to extract the main features of an event.

Charged particles can travel over 2 metres in the central detector, and yet each measured point is measured to an accuracy of a few hundred microns. When the whole event is displayed, it is not possible to observe small deviations or inconsistencies of the hits belonging to a track. However, the display can be adapted to stretch the picture in one direction and to allow careful visual inspection of fine details of the reconstructed track, which during the experiment was a crucial aid to the verification of central detector tracks where a high momentum had been reconstructed. Any quantities that have been computed for the event can be added to the event display, including the reconstructed jet directions and energies, with their component tracks or cells. When a 'neutrino' has been deduced from the transverse energy imbalance observed in the calorimeters, this can be displayed and inspected to check if this is a reasonable interpretation. When two or more particles are suspected to result from the decay of a short-lived particle, then the mass of the parent particle can be calculated interactively.

These event displays were used around the clock in the search for the W and Z bosons and all selected events were studied briefly (scanned) on these display screens. The original aim of the scan was to identify any problems with the detector and to reject any events which were clearly invalid. Once this was done the accepted events were classified into different categories and general features of these events were recorded on scanning sheets together with copies of the event 'picture' for future reference. Unusual events generated great excitement and were studied carefully by experts from all parts of the detector. For the small number of events which are of most interest, a further analysis stage is used. The event is displayed again and all aspects of the event reconstruction are checked carefully. In particular the reconstructed track is subjected to a detailed analysis of the effects of changing the central detector points along the track. In this stage, known as fixup, the track can be refitted several times with slightly different points, in order to understand the sensitivity of the results, for the track momentum.

These graphics displays were used at CERN and many outside laboratories as an important aid in the search for the W and Z bosons. All interesting events were closely scrutinised and the scanning area, like the experimental control room, was a focal point for physicists during and after the experimental run. It is very much easier to explain and understand the results of an experiment using such a clear event display. Although the particles under study are so small, they leave clear evidence of their presence in the detector. When you have spent many hours examining the antiproton–proton collisions on display screens, it is much easier to visualise these tiny particles as their paths are reconstructed in front of you, in a way that you can see! This complex transformation, from recorded raw data to a fully reconstructed collision, was achieved by the efforts of many people. All the programs were developed and tested in time for the start of the experiment.

We have now outlined the variety of computers involved in the UA1 experiment which were all utilised around the clock. As this collaboration included 150 people, there were usually several using each of the computers at all hours of the day or night. Whatever time of the night, there were always a few stalwarts trying to finish some urgent piece of computing. On one afternoon I started using the main CERN computer from Birmingham University and noticed to my astonishment that there was not a single other user from the UA1 experiment on the machine. I soon heard the reason – the first Z^0 candidate had been found and all the UA1 experimenters at CERN were jammed into a conference room discussing its properties.

9.6 The physicists and the shifts

The list of people 'without whom the experiment would not have been possible' is very long and any attempt to be comprehensive would undoubtedly miss out some very deserving cases. My intention here is only to give you some idea of the way in which this large collaboration operated. It should be stressed that the responsibility for the accelerators and the very successful machine performance rested with the CERN staff. The various groups of accelerator physicists, engineers and operations staff responsible for the successful PS, SPS and antiproton accumulator (AA) operation deserve great credit. Without this success, all the experimenters' careful preparation would have been in vain.

Let us now examine the various tasks performed by the experimenters in the UA1 experiment and build up a picture of their contributions. Although this description follows that of the accelerators, detectors and computers the experimenters make an even more important contribution than any of the previous items. The technicians, engineers and physicists provided years of conscientious effort to the project. The routine tasks are usually the most time consuming and in these the computers are a great help. However, the installation and systematic checking of the mechanical and electrical components of such a large detector requires tremendous effort, care and patience from a vast number of people, and could not be automated far beyond present levels. Good communications and organisation are also essential in such a large venture. However, the unique resource brought to the project by the many experimenters is quite different. They contribute a series of new ideas for improving the detector and the selection of rare events. In addition they carefully validate the many features of recorded events and use their intuition and knowledge to suggest reasons for problems in the detector or account for experimental surprises. This flair, when added to the total commitment of the experimenters to the success of the project, makes the personal contributions absolutely essential and leads to a very exciting working atmosphere.

Any large experiment, which records data over many months, needs a solid core of people living near Geneva. In the UA1 experiment the CERN group and a fraction of the physicists from the other collaborating groups were based for long periods at CERN. Some of the collaborating groups were from research institutes, whereas others were from universities, where staff also contributed to the teaching in their departments. Most of the groups contained a range of personnel including permanent staff, those with temporary appointments and students studying for a higher degree in physics. The resident experimenters usually have apartments in Geneva or the neighbouring towns or villages of France and Switzerland. Some members of the collaboration, including myself, travelled regularly back and forward to the experiment at CERN, from their own university or research laboratory. This group are called commuters, even though the distances travelled are rather larger than most of us would choose, and these visitors frequently stay in accommodation at CERN. In this way the problems of transport at irregular hours are avoided and it is much easier to keep closely involved with the experiment and its analysis. All members of the collaboration participated in the data-taking runs and were also involved in some aspects of the detector construction and event analysis.

We shall be concentrating on the running periods of the experiments later in this book; however, it is worth recapping on the variety of work performed by experimenters in such a large project. In the design stage, which was completed over several months in 1978, people from the various institutes worked together to arrive at an agreed proposal. When this was accepted the responsibilities for constructing and testing each piece of the detector were allocated to the collaborating groups in a way that was consistent with their financial and manpower resources. At this stage each group has its own responsibilities to be completed and so careful coordination was essential to make sure that the various pieces would fit together!

The experiment was still several years from taking its first data but there were many computer programs to be written, more than a million lines of computer code, so an early start was essential and the most important computer programs were written, developed and tested. As the various components of the detector were constructed their quality needed careful checking and they all had to be fully tested and calibrated. Gradually the elements of the detector arrived at CERN for further testing and assembly and the pressure increased as the critical deadlines became closer. In the period before the data-taking run, the collaboration constantly reviews its programme and tries to move extra effort to any parts of the project which are behind schedule. Once the construction of the detector is complete, the responsibilities are shared and a closer collaboration develops. In the initial phase most people contributed to the design and testing of components, but gradually this changed as more and more people became involved in the computer programs used for the checking and analysis of the data.

At the time of the run the apparatus has to be monitored 24 hours a day, seven days a week, and to help to make the coverage of people more uniform a shift system is operated. The various parts of the detector vary in complexity and some require very experienced personnel who can be contacted by a remote paging system at all times. During most parts of the experimental run, there would be eight people in the experimental control room. Individuals would have responsibility for monitoring the data acquisition system, electronic trigger, central detector, calorimeters and muon chambers and one of these would also perform routine safety checks on the gas supplies and other pieces of the detector. One person would be running the monitoring programs, another performing the rapid analysis of the express line tapes and another scanning recently recorded collisions at a display terminal. Finally, a shift coordinator in charge of the operation would be responsible for communications and take responsibility for any changes in running conditions that were necessary. This team would work for an 8-hour shift changing at eight in the morning, four in the afternoon and midnight and although each person had a similar number of shifts, these would occur at frequencies which were different for residents and commuters. Typically a resident would have three shifts a week while a commuter would do more shifts to make up for the periods spent away from CERN.

These shift periods are the basic minimum that each physicist needs to attend to be a member of the collaboration. However, the vast majority of the collaboration spend so many hours working on the experiment, that these shift periods represent only a small fraction of the total time. In the early phase of an experimental run there are usually numerous mysteries and problems to be solved and anyone who has the expertise to help usually stays working on the problem until it is solved, whatever the time of day or night. There are plenty of tired physicists around during this phase of an experiment.

The official scientific language at CERN is English and this is used for most of the meetings, seminars and documents. Many of the participants do not speak several languages fluently and so there can be misunderstandings caused by language alone. However, there is a genuine friendship between the many participants and it is very satisfying to be a member of such a cooperative enterprise. Members of the collaboration are usually far too busy to be unduly concerned about anything other than getting the experiment running successfully. Much of the credit for the success of the collider experiments must go to the many physicists in the collaboration. At all hours, all over CERN, they could be found checking the equipment and analysing the data. Without their conscientious efforts the W and Z bosons would still be predictions.

Part Four

The search

Chapter Ten

Installation and the early runs

10.1 The underground areas are ready

We have now described the targets of the search and all of the experimental preparation that was needed for it. We have also outlined the changes to the CERN accelerators that took place in preparation for the first antiproton–proton collider run. The three underground experiments, UA1, UA2 and UA5, have been constructed and in the summer of 1981 the search for W and Z bosons was about to start. The experiments UA2 and UA5 are to share an intersection many kilometres from CERN and over 60 metres underground. It was planned that UA5 would take data in the early, low-intensity runs, to be replaced by the UA2 experiment for all of the later high-intensity running periods.

The UA1 experiment was being installed in an underground area about 1 kilometre from CERN, just a few hundred metres from the outskirts of Geneva and very close to the French–Swiss border. Two separated 20-metre diameter holes had been drilled to a depth of 30 metres, one directly above the super proton synchrotron (SPS) accelerator tunnel, and the other just inside its circumference. These were joined near the base to allow the UA1 detector to be moved from one to the other at data-taking periods, but between runs the detector was located in the pit, known as the garage. For a short time this area looked like a typical muddy construction site (figure 10.1*a*,*b*) but this rapidly changed. The top of the pit above the accelerator tunnel was covered and has now reverted to green fields whereas the top of the garage is covered by a two-storey building. The outer shell of the building protected the pit and the assembly areas at the surface from the weather and one corner of the building was partitioned off

(a)

(b)

Figure 10.1. (*a*) An early view of the underground area which was later occupied by the UA1 detector. (*b*) A later view before the hole above the super proton synchrotron (SPS) tunnel was covered. In the centre of the picture the wall that separates the 'garage' from the SPS tunnel can be clearly seen. (*c*) The first element of the UA1 calorimeter being carefully lowered down to the underground area. Photographs courtesy of CERN.

into small rooms, which were used as electronic workshops. The experimental control room was built in the same area on the first floor, above a room of computer terminals and a small room that was to be the scene of many lively meetings.

10.2 The UA1 detector is assembled

It was time for the UA1 detector to be fully assembled, and as the detector weighed over 2000 tonnes and it had to be assembled at the bottom of the pit, this was a major project. Most pieces of the detector had been constructed in modular form to make this job easier, but every single large piece of the detector had to be lowered, using massive 60-tonne cranes, to the base of the pit for assembly. The scene at the empty experimental area, when the first element of the hadron calorimeter was lowered, is shown in figure 10.1 *c*. Once it was assembled the whole detector needed to be mobile so that it could be moved through the connecting passage into the SPS accelerator tunnel for the data-taking periods.

The whole detector was constructed on a large wheeled 'chariot' supported on rails, and when the detector was complete the whole assembly could be moved. However, once the detector was closed,

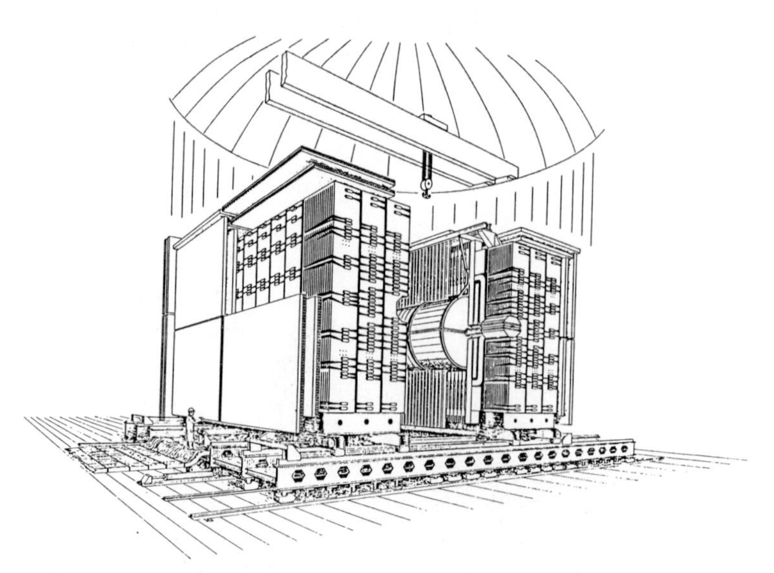

Figure 10.2. A schematic view of the entire UA1 detector before it is closed around the central detector. The two sides are closed up before the whole detector is moved into the super proton synchroton tunnel. Some of the outer planes of muon chambers are shown surrounding the hadron calorimeter.

Figure 10.3. A picture of the assembled UA1 detector before the muon chambers were added. Photograph courtesy of CERN.

it was very difficult to access many of the elements and so the two separated halves of the calorimeters were first assembled (figure 10.2). Once these had been checked the two sides of the calorimeter were then moved together to close around the central detector (figure 10.3).

All of the elements of the detector were checked before they were lowered to the underground area. However, as it was very difficult to move these large items delicately, there was often further work to be done at the bottom of the pit before final assembly. It is very difficult to convey in words the scale of activities that is needed to complete the installation of such a large detector. As many of the physicists, engineers and technicians as possible are involved in the last-minute checks of the apparatus and there are also many other temporary staff hired by CERN to help at this time. This methodical and often very tedious work lasted for a period of many months with every single cable being cross-checked to ensure that each piece of the detector was connected up correctly. Problems which were not isolated and dealt with at this stage could not usually be solved during the run, because there was only limited access time to the detector. The care and accuracy taken to perform these thousands of systematic checks were crucial contributions to the successful operation of this complex detector.

The view from the top of the pit is very impressive at any time (figure 10.4), but in the days before data taking there were people clambering all over the detector. There is only one lift to this area, but in addition there are ladders around the sides, which wind their way down to the bottom. Fortunately there were no serious accidents, but with so many people and so much activity, great care had to be taken and for people working in the pit, builders' hard hats were an essential item. The detector is over 10 metres in height and so it is difficult to reach even the outermost elements in some cases. Long ladders and climbing harnesses were essential at this stage and for those people balanced near the top of the detector three hands would have been very useful!

The information from the experiment is carried out from the apparatus by tens of thousands of wires and so the overall weight of wire is enormous and large cable supports are required. These supports need to be flexible because the detector has to be moved along the rails taking it from the garage shaft, where all the preparatory work is done, into the main shaft where it is placed in the SPS accelerator tunnel. As the detector weighs 2000 tonnes, this movement requires careful organisation. Once the apparatus has been rolled into place in the acclerator tunnel, the access between the shafts is closed, by moving large concrete blocks into place. After this, the only way into the experiment is through the special interlock doors, which allow access to the accelerator tunnel at a few locations. These cannot be opened while the accelerator is running, and so the detector can be reached only for very limited periods, when both experiments decide on an interruption to the run. Any person entering the accelerator tunnel needs an identity card to open the door and as a safety precaution is required to remove a separate key. The accelerator cannot be restarted until all the keys are returned and so people return them very promptly!

The cables from the detector are taken to a two-storey cabin, known as the MEC, for connection to the electronic triggers and other electronic equipment. This is also at the base of the pit and moves when the detector is rolled into the accelerator tunnel, but the MEC remains on the outside of the concrete wall and is accessible

Figure 10.4. A picture of the partially assembled UA1 detector from the top of the pit. The two halves of central calorimeter are separated in readiness for the installation of the central detector. Photograph courtesy of CERN.

during the run. It consists of two rooms, one above the other, each about 10 by 3 metres and these are crammed with electronics and cables which occupy over half its volume. In addition there are computer terminals and communication systems to the experimental control room at the surface. The racks of electronics need to be cooled and the operation of the fans make it a very noisy place in which to work. In fact, as there are several groups of people needing to work in these rooms simultaneously, it can get very difficult to fit everybody in. Cables from the MEC carry the measurements from the detector to the data acquisition computer at the surface. Most events are rejected as uninteresting by the electronic triggers and so these never reach ground level!

The logistics of installing and keeping a record of tens of thousands of cables should not be underestimated. Months of work is required to ensure that each cable really comes from the expected wire or calorimeter cell, and without this thorough preparation the later analysis would be impossible. This checking is often tedious because it usually involves sending a test pulse from each wire or photomultiplier and checking that it arrives at the expected location. Lasers are used to send light down long thin fibres into every single photomultiplier to check that the electrical signal is received in the correct cable. Each wire of the central detector and muon chambers, more than 6000 of each, are activated with test pulses and their response checked. Although computers are very useful in this sort of work, physicists and support staff devote many hours to this important task.

10.3 The first collisions

In February 1981, the proton synchrotron (PS) at CERN was used for the first time to accelerate antiprotons which had been stored in the antiproton accumulator (AA). In June of the same year the SPS accelerator became operational again, after a shutdown of one year for essential modifications for the antiproton–proton collider programme. It would clearly be some time before there was a chance to run with high-intensity antiproton beams. Would the complex collider system work at all? Much effort had gone into preparing the detector for the earliest possible period of data taking, but for these first tests it was decided only to install the trigger scintillators on each side of the intersection, which could be used to tell whether any antiproton–proton collisions were occurring.

In the first month of operation the SPS was used to accelerate protons as it had before the shutdown, but on 7 July 1981 it was used to accelerate its first bunch of antiprotons to 270 GeV. Two days later this was repeated with both antiprotons and protons in the SPS accelerator. At this stage both beam intensities were very low compared to later running periods and the proton bunch contained many more particles than the corresponding antiproton bunch. However, there was great excitement about the prospects of observing the first antiproton–proton collisions at 540 GeV, even though at low intensities the background beam gas interactions were expected to dominate the real antiproton–proton collisions. Genuine antiproton–proton collisions can be distinguished from background beam gas interactions because signals in the scintillation counters at both the front and back of the detector should be observed at the same time.

At the start of the short run, beam gas collisions from each side of the detector were observed and then gradually, as the number of events increased, the first antiproton–proton collisions ever observed at 540 GeV became evident. A small fraction of the events showed both scintillation counters detecting particles at the same time, proving that antiproton–proton collisions were occurring at the expected intersection point. A typical early distribution of the time difference between the scintillation counter signals at each end of the detector is shown in figure 10.5. The two outer peaks indicate the presence of beam gas collisions and the central peak corresponds to antiproton–proton collisions. The clear separation between the peaks emphasises the clean distinction between the background events and the genuine collisions. Calculations were quickly made to compare the observed number of collisions with the expectations for the known beam intensities. The results looked consistent – at last the antiproton–proton collider was operational. Carlo Rubbia delayed his departure to the International Conference on High Energy Physics in Lisbon while these measurements were made. On 10 July

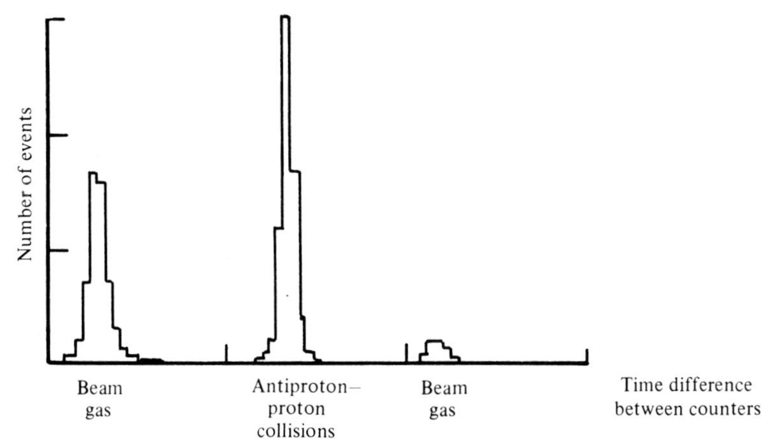

Figure 10.5. A typical early distribution for the time difference between signals detected at the scintillation counters at each end of the UA1 detector. The central peak, where the time difference is small, corresponds to genuine antiproton–proton collisions. The other two peaks, where there is a time difference between the signals, indicate the extent of beam gas collisions. The larger peak corresponds to proton beam gas collisions as there are more protons than antiprotons in the beams.

he announced at Lisbon that the first antiproton–proton collisions at 540 GeV had been detected at CERN.

During these early tests of the accelerator, development work was performed with proton beams, in order to minimise the losses of the precious antiproton beams that were still of very low intensity. Much more work was still needed by the accelerator experts and operational crews to develop a reliable way of transferring the particle beams between the various accelerators. We have already described the complicated sequence of beam transfers that were necessary and each of these stages needed to work efficiently and consistently before there was any hope of starting the main search for W and Z bosons. In particular, the timing of all the transfers between these accelerators had to be very carefully synchronised, otherwise the beams were lost. This meant that the beam collided with a magnet or beam pipe, or was gradually defocussed and deposited around the accelerator. The accelerator is well shielded with earth and concrete to avoid such losses causing any external problems, but it usually caused a serious delay to the experimental run when the beam was lost. Once a beam transfer failed there was naturally a certain reluctance to repeat the operation until the reasons for the problem had been understood. Consequently there was usually a long sequence of 'pilot shots' where only a small number of antiprotons were transferred, and only if these were successful was another attempt made to transfer a 'big shot'. A further loss in this second transfer could mean a wait of up to a day to refill the antiproton accumulator with a reasonable number of antiprotons.

The UA1 collaboration was very keen to insert at least part of its detector into position in the accelerator tunnel at the earliest opportunity. This would enable an early start to the study of these highly energetic collisions and also provide useful experience of operating the detector in realistic experimental conditions. However, while the control of the beams was still being improved, there were serious risks that the intense proton beam could damage the carefully constructed detector, especially the delicate central detector. As a compromise it was decided to install the UA1 detector without this central track chamber in August 1981. This was just the second serious attempt to achieve antiproton–proton collisions and so it was a period mainly for machine development rather than a genuine data-taking run. In August 1981 the number of antiprotons in a bunch reached 10^9 for the first time, still a factor of 100 below the design value. Another important property of the beam is its lifetime, which even in these early runs was already as long as 20 hours, close to the design value. This run was very exciting as it was the first chance to study the properties of these high-energy collisions at an accelerator and it was the first chance to see whether the calorimeters would perform as expected. There had been tantalising hints from cosmic ray experiments that some very unusual reactions would be observed at such high energy. One feature from the earliest studies of this run was the large differences in the energies detected in the two ends of the detector due to beam gas collisions. There were many more protons than antiprotons and so the beam gas collisions occurred predominantly on the side of the proton beam. Once these events were excluded by the pretrigger, over 4000 examples of genuine antiproton–proton collisions remained. These events looked symmetric and from a physics viewpoint were disappointingly normal with no big surprises. However, most important parts of the detector had now been successfully tested in operational conditions, together with the analysis programs.

For the periods of normal proton running at this time the UA1 detector was opened up and left inside the accelerator tunnel. This was to reduce the time spent in moving the detector to and fro between the garage and the accelerator. During one of these runs the calorimeters at each end of the detector were repeatedly struck by the intense proton beam, which caused radiation damage to the sheets of scintillator in a limited area of the detector. In a few cells this reduced the scintillator light output quite considerably, but fortunately the damage was not too serious and these cells were recalibrated. The very early collider data recorded by the UA1 experiment had been achieved at the cost of some damage to part of its calorimeter.

The next short run was scheduled for October 1981 and the UA1 experiment was about to make a major step forward with the installation of its central detector, which had by now been constructed and tested. This was a very advanced detector, using a more adventurous design than any other drift chamber of its size, and it simply had to work for the experiment to have a chance of success. Many people did not believe that such a device could be designed, constructed, tested and installed in less than 3 years. The mechanical part of the detector was ready, although only a small fraction of the advanced electronics that were needed to read out the information from the 6000 wires was available at this time. These electronics were essential to record the passage of all charged particles detected by each wire. When many particles are produced close to each other, the electronic signals which the particles generate reach a wire at very similar times. The fast electronics needs to be able to separate the arrival times of these signals and even measure the deposited charge due to each particle. This was a period of major uncertainty. Would this complex detector operate reliably and accurately when hundreds of charged particles passed through it? There had been serious delays in the delivery of electronics and it was not clear whether there would be enough to allow the use of the central detector in this run. In fact, the first deliveries arrived only weeks before the run and the electronics were being installed right up to the last minute.

This data-taking run started with the central detector installed and connected to all the electronics that had been delivered. It had been decided to read out every second wire in the central part of the track chamber because even this required electronics for over 2000 wires and was a very big advance on the previous run. The magnetic field of the UA1 detector was still switched off in these early days to minimise any possible complication to the circulating beams, and so the momenta of particles could not be measured.

The antiproton and proton beams were injected and accelerated to 270 GeV and were kept circulating at this energy. Once the beams were stable, the high voltage on the central detector was gradually raised towards an operational value. The colour monitor in the experimental control room was used to display the hits recorded by the central detector, and once the voltage was high enough these hits appeared. On each event many charged particles could be seen travelling out from a common vertex, and so at last the full UA1 detector was operational, and recording the properties of the energetic antiproton–proton collisions. Even with limited electronics the central detector produced a beautifully clear visual record of each interaction (figure 10.6).

10.4 The first results

The first tapes of these data were hurried to the central computers and displayed on the high-precision graphic displays. At first, the scanning concentrated on identifying problems with the detector, but then quickly switched to measuring the properties of these collisions which had never been observed before. How many particles were produced and how were they distributed? These and many other properties were rapidly determined and a first publication was immediately prepared, summarising these early results.

At the other intersection, the UA5 streamer chamber was also working well, recording the first photographs of antiproton–proton collisions. These photographs were rapidly developed and scanned at CERN and the collaborating laboratories. This was performed on scanning machines, where the image on the film is magnified and projected onto a horizontal surface where it can be moved to allow

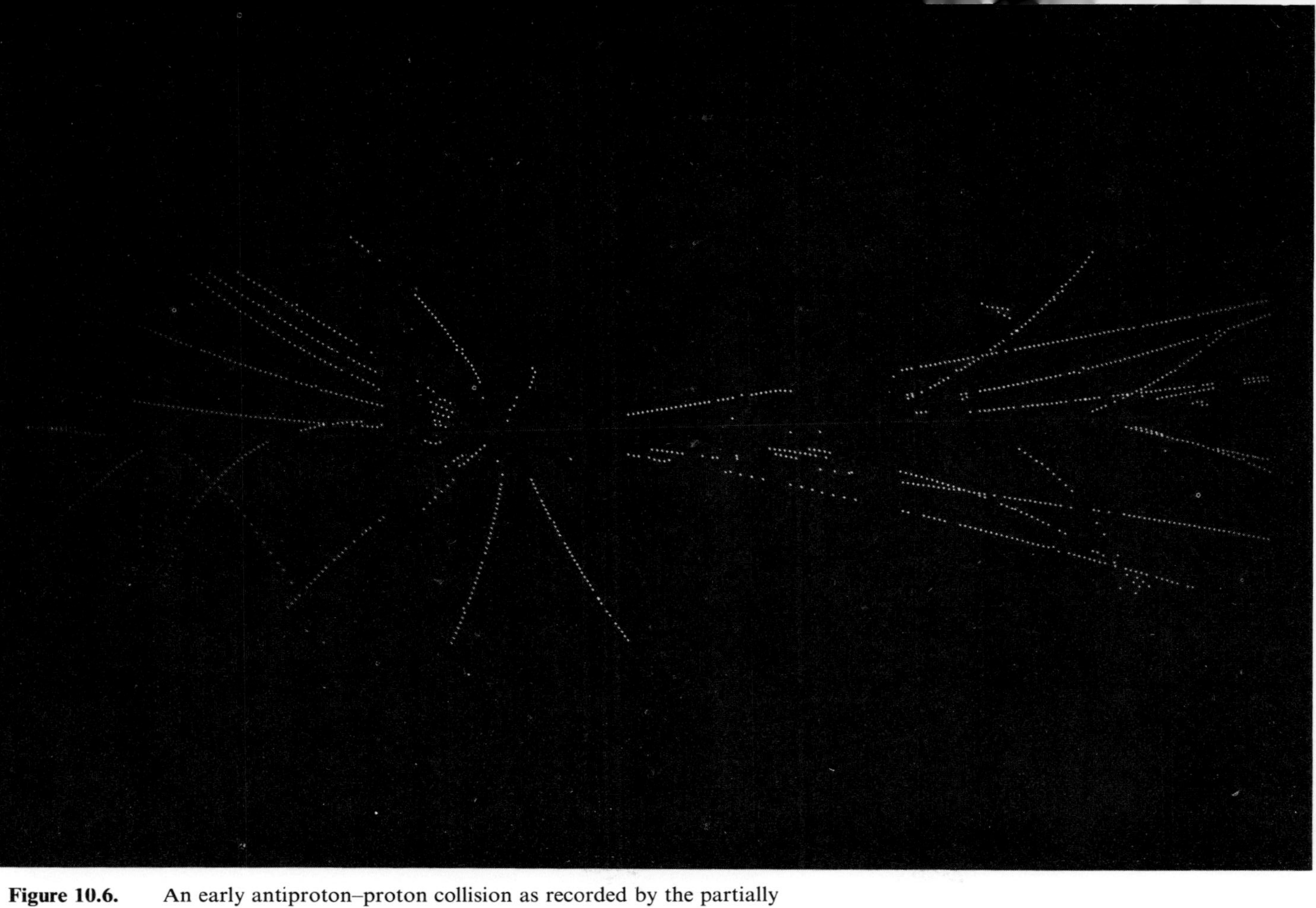

Figure 10.6. An early antiproton–proton collision as recorded by the partially instrumented UA1 central detector. This computer display, of the information recorded by the central detector, clearly shows that a collision has occurred.

closer inspection of any part of the picture. The UA5 collaboration also rapidly scanned and measured these photographs of antiproton–proton collisions and published these results. The first results on these energetic collisions, from two very different detectors, were very similar which provided a further check that the detectors and subsequent analysis procedures were well understood.

These early data were rather similar to those observed in lower-energy experiments, and although the number of particles produced was higher, it was clear that no dramatic change had occurred. The anomalous results recorded by cosmic ray experiments at similar energies disappointingly had not appeared. However, it was also encouraging news because the estimates of W and Z boson production rates and of the possible backgrounds had assumed a smooth variation between the low- and high-energy experiments. The antiproton beam intensity was still far too low for there to be a chance of discovering the W and Z bosons, but these early periods of data taking were an ideal way to 'run in' the detector and the analysis programs, and also to work out the best strategy for the later runs.

For the final collider run of 1981, the UA5 streamer chamber was replaced by the UA2 detector and so both the major experiments that were designed to observe the W and Z bosons were now in position and ready to take data. More electronics for the readout of the UA1 central detector had arrived, and hopes were high that a long period of data taking would be possible, although it was not expected that the beam intensities would be high enough for the W and Z bosons to be identified. The UA1 magnetic field was switched on for the first time in this run, so that the momenta of charged particles could at last be measured.

From the early days of the run the detectors worked very well but there were problems with the accelerator complex. It took a long time to develop reliable techniques to produce, transfer and keep the proton and antiproton beams in the various accelerators. The instrumentation for monitoring the beams was rather limited and so it was difficult to be certain what was causing the problems but there were often long delays between the loss of a beam and the installation of the next 'shot' of particles. This was frustrating both for the experimenters waiting patiently on their shift and the operational crews of the various accelerators, who were doing their best to deliver reliable beams of antiprotons and protons. Many lessons were learned in these early months but it was a slow painful process.

The experimental data taking was interspersed between many periods of accelerator testing. In total, only 40 hours of collider physics had been obtained before the 2-month run was terminated shortly before Christmas 1981. Only modest intensities were achieved and the highest intensity lasted for just a few hours. Nevertheless, hundreds of thousands of events were recorded, and this was the first extensive data taken with both major detectors close to their final configuration. Although the intensities were too small for W and Z boson production there were many other phenomena to study at this very high energy of 540 GeV, and shortly after the run both experiments analysed their data rapidly. This analysis was performed in parallel with preparations for the next experimental run, which was planned for the spring of 1982. In some of the collisions, the particles were emitted in close groups or jets, reminiscent of the jets that had already been observed at electron–positron colliders. These fragments of quarks and gluons were clearly visible for the first time in hadronic collisions and some of these events were very dramatic, with the jets carrying transverse energies as large as 100 GeV.

The data-taking run at the end of 1981 had provided some very interesting new experimental data. In addition, it had been an extended opportunity for operating the full detectors and improving the techniques for rapid conversion of the raw data into interesting physics results. However, there was still a major uncertainty over the central problem. Could the beam intensities of protons and antiprotons be increased by the large factor still necessary to give the experimenters a chance to find the W and Z bosons? Enormous improvements in both the maximum intensity and the overall consistency of collider operation were necessary for a realistic experimental search for the bosons. Many lessons had been learned and as the next collider run was scheduled for a few months time perhaps it would provide the first chance to detect the W and Z bosons.

Chapter Eleven

The search for the W boson

11.1 The delays to the run

It had been planned to continue with collider physics in the spring of 1982, after the winter shutdown of CERN, but shortly before the running period was due to start, there was an accident involving the UA1 central detector. The delicate detector, which is packed with electronics, was being cleaned with compressed air in readiness for installation at the intersection. Unfortunately the air line was connected to a faulty supply line which contained dirty water and the delicate components were contaminated. There was no alternative, the whole central detector had to be lifted out of the pit by a crane and taken back to its preparation area and completely dismantled. Each component needed to be carefully cleaned, tested and reassembled. This would take many months, and as the accident followed the rather disappointing beam intensities of the 1981 data-taking run, it was probably the lowest point of the whole collider programme for the UA1 collaboration.

Following this accident, the CERN management decided to delay the 1982 collider running to a single long period at the end of the year, starting in October. This had several advantages because it allowed both experiments a breathing space to prepare the detectors very carefully for the next important run, but at the time this additional delay was very frustrating for everyone. A single long collider period was also more convenient for the accelerator staff, as they could concentrate on one type of accelerator operation for several months, rather than frequently changing between very different operating conditions.

In the summer of 1982 another accident occurred at the UA1 underground area. During one of the frequent thunderstorms that occur through the summer, water poured down into the underground area faster than any pumps could remove it. The water level rose rapidly and it reached the lowest part of the detector before the pumps managed to keep the water at this level. Later that day the water was pumped away but the muon chambers, which are at the base of the detector, were seriously damaged by the water. There had been careful tests of each chamber, in test beams at CERN, in order to check their operation. These drift tubes operate at very high voltage and each module includes hundreds of electronic amplifiers which had been subjected to comprehensive tests. This tedious work had only recently been completed, so you can imagine the reaction when people realised that much of this patient work would have to be repeated. However, there was no alternative, the muon chambers were rapidly removed from the base of the detector and craned to the surface of the pit, where a crash programme was started to remedy the damage caused by the water. Each of the damaged muon chambers had to be opened and all of its mechanical and electrical components carefully cleaned. Although there was still some time before the next run, the checks of the muon chambers needed to be finished very urgently. The detector assembly is a very complicated sequence of operations and these muon chambers need to be installed before the other items. As there was no time to take the chambers back into test beams, the ever-present cosmic rays were used to test the chambers. This round the clock work was completed just in time for the chambers to be reinstalled at the base of the UA1 detector.

In the summer of 1982 spirits were at last being raised. At the major particle physics conference in Paris, the results from the collider running of the previous year were receiving wide attention. The highlight was certainly the observation at the collider of dramatic jets of energetic particles, produced for the first time in antiproton–proton collisions. This type of event had been recorded by UA1 and UA2 detectors and both collaborations had analysed their data very carefully. However, the UA2 collaboration presented the most dramatic results on jets to the Paris conference and attracted most of the attention. This provided a timely reminder that this search was very competitive and that neither collaboration could afford to relax if it wanted to be the first to detect the W and Z bosons. Both collaborations displayed events where enormous energy was concentrated in limited regions of the detectors. This had never been observed as clearly before in any collisions between hadrons. As the proton and antiproton are made up of quarksa, antiquarks and gluons it was expected that in some collisions a very hard scattering would occur between these constituents. As the scattered quark or gluon cannot exist in isolation, it converts itself into observable particles and these form a jet. In electron–positron collisions these jets had frequently been observed but it was very exciting to record a similar phenomenon in the more complicated collision between a proton and an antiproton. Further analysis revealed that these events could provide very useful information on the internal structure of the proton and the forces between the quarks. In fact the scattering distribution of these quarks and gluons is remarkably similar to that first observed by Rutherford in his studies of α particles scattered from the nucleus. This similarity indicates that the interquark force also follows an inverse square law, when the distance between the interacting quarks is very small.

11.2 Last-minute preparations

As we will soon be describing the search for the W boson, let us review a few important features of this particle. It is expected to decay to an electron and antineutrino, which can each carry up to 40 GeV of transverse momentum. The energetic electron is identified by a straight track in the central detector, followed by a 40 GeV energy deposit in the early samplings of the electromagnetic calorimeter. The presence of the energetic antineutrino is deduced indirectly by assuming momentum conservation in the transverse plane. The W boson is expected to be produced once in every 100 million proton–antiproton collisions. To achieve this number of collisions it is necessary for the intensity of the antiproton beam to be increased very significantly over the peak 1981 value and for the beams to be maintained at high intensity for long periods.

The period before the 1982 run was even more hectic than usual but as it was the first real chance of discovering the W boson, this added an extra incentive to the preparations. The UA1 and UA2 detectors were both designed to observe the W and Z bosons and so there was a strong element of competition between these experiments. As usual the final checks to the detectors were made right up to the beginning of the run. While these checks of the detector are in progress, other people are making last-minute improvements or corrections to the many computer programs that will soon be needed. As the run approaches, the experimenters spend more and more time at the experimental area. The conference room, directly under the experimental control room, is used for daily collaboration meetings. Recent progress, or lack of it, is discussed and there are also many smaller informal discussions on the detector and analysis strategy.

The single coffee machine in this area begins to increase its sales by an enormous factor and change for this becomes the next most important item to the identity card which is needed to enter the building. During the run an enormous box of small change accumulates in the control room so that people can always obtain a coffee to help them through the night.

The meetings would vary enormously, some were brief and others would drag on interminably. Many were conducted with good humour but others were very depressing. Sometimes elements of equipment (hardware) or computer programs (software) were not ready at the necessary time. At other times there would be a conflict of interest between different members of the collaboration and sometimes poor communications would magnify a comparatively trivial problem into a more serious one. Tensions are usually greatest at the start of a running period, as there are so many uncertainties and things to be finished. After a period the experimental run becomes more routine and the pressures switch to the early analysis of the data. As the pressure to meet deadlines gets more severe and the working hours get longer and longer, there are bound to be problems. Happily these were rather few and far between, as the cooperation was very good between the vast majority of the collaborators. The coordination of this vast exercise was another major responsibility which was handled by the spokesman of the experiment, Carlo Rubbia, from CERN, and his deputy, Alan Astbury, from the Rutherford Appleton Laboratory. During running periods one of these two, usually both, would be present at the daily meetings at the experimental area and they also acted as contact people with the accelerator staff at CERN for the collaboration. The 150 experienced physicists were individually highly self-motivated. The lively personality of Carlo Rubbia was able to channel efficiently this enthusiasm and provide even more motivation at some crucial stages of the experiment, but ultimately this was a very impressive team effort.

At the start of the run the beam conditions are usually very poor, and there are many technical problems to be resolved, which means that many people stay at the experiment for very long hours trying to solve the problems. One evening, at the beginning of the 1982 run there were so many people in the control room that it became very difficult to run the experiment. On this occasion people who were not officially on shift were told to leave, but this was a very unusual occurrence as usually any extra help was gratefully received.

11.3 At last the run starts

There was the usual hectic start to the run in September 1982. While the accelerator staff prepared for the long experimental run, the antiprotons were being stored and cooled in the antiproton accumulator (AA) and last-minute calibrations and checks were also being completed at the experiments. A new 'shot' of protons and antiprotons needs to be injected into the super proton synchrotron (SPS) on average once every day and as the antiprotons have taken many hours to collect this is always a critical operation, even towards the end of a run. The accelerator crew who are responsible for the delicate operation usually try a few 'pilot' shots with a small number of antiprotons to check and optimise the transfer. At the experimental control room a microprocessor generates an electronic countdown, with a Japanese accent, at each of these shots and for a moment it seems like mission control at the Houston Space Centre. Finally the big shot of antiprotons is transferred into the SPS, rapidly followed by the injection of a proton beam, and their orbits are checked carefully. Once the three bunches of protons and antiprotons are circulating in a stable manner the detector magnets can be powered. The UA1 magnet requires a current of 10000 amperes and at either side of the main magnet the beam passes through smaller magnets which correct the deflection of the circulating beam.

Data taking is started at the earliest opportunity because the beam intensities are highest at the start of each shot and so it is most important to use this period efficiently. In normal running one shot lasts for a day, but there are many exceptions because sometimes the beams are lost through a power failure or other electrical failure. Occasionally the beam intensities decrease more rapidly than usual and then it is advantageous to refill with new beams, providing enough new antiprotons have been stored. It usually takes several hours to refill the SPS accelerator and so any decision to deliberately 'kill' the beam is not taken lightly. If this decision is taken, the circulating protons are ejected using 'kicker' magnets onto a specially prepared metal dump, whereas the less-intense antiproton beam can be safely distributed around the accelerator tunnel.

Once the data taking is started, a magnetic tape is filled with information once every 10 minutes and, as described earlier, emulators are also used to perform an even tighter selection than the electronic trigger. When an event passes these extra selections, it is a more serious candidate for a W or Z boson and it is immediately written onto an 'express line' tape, which is filled about once an hour. These tapes are rapidly copied and taken to the main computer centre where the analysis programs are used to reconstruct fully the events. The raw recorded data from each element of the detector have to be converted into calibrated measurements. These are then linked together to build up a very detailed 'picture' of the properties of particles produced in each collision, including their directions and energies. Finally, each collision is analysed to search for evidence of any energetic electrons, muons or neutrinos.

The most interesting events are then selected for inspection by the scanning shift. Occasionally events look so dramatic from the computer output that a quick phone call is made to the scanning area to ask for an immediate check of a particular event. After careful scanning and checking some of the selected events are rejected. There are many different reasons for this, for example sometimes the dramatic energies are deposited by cosmic rays or other particles which strike the detector from the outside. The analysis programs can sometimes be confused by very complicated events or where one region of the detector is particularly active. The events with obvious problems are rapidly eliminated, but many genuine events remain for more intensive study.

Back in the experimental control room many things are also happening as the routine monitoring of each piece of the detector continues and details of beam and detector conditions are entered into numerous log books. At the end of the run these detailed descriptions, which are continuously recorded, are usually the only sources for reconstructing the exact conditions of the run. Every few hours some, usually minor, problem with the detector or data acquisition computers would occur. Sometimes the amplifiers on the muon chambers would start to oscillate and produce many electrical signals in the detector, generating major problems for the data acquisition system. The remedy was usually very simple – just switching off the power to these amplifiers for 30 seconds cured the problem but in the early days this was quite an adventure. When the oscillations started, a signal was displayed on an oscilloscope, near the person responsible for the muon shift. If this was spotted, usually accompanied by a general shout from colleagues, it was necessary to race down to the bottom of the pit as fast as possible to switch off the amplifiers. In the later running this hectic scramble down the stairs was no longer necessary as the power switches were installed in the control room!

Similar incidents livened up the shifts for the other parts of the detector and the expertise of the people on the shifts varied enormously. Some of the experts had played key roles in the construction and testing of pieces of the detector and were able to diagnose problems very quickly. Similarly with the data acquisition system, some people had written many of the programs used and had been intimately involved with these computers from the beginning. These hardware and software experts played a key role especially during the early part of a run. For several weeks they would be at the experimental control room at all hours trying to ensure that no beam time was wasted. The other people on shift performed the usual monitoring and recording tasks in data taking periods, but when unusual problems occurred it was usually necessary to call in an expert.

The coordinator in charge of the shift has to monitor the beam intensity and how many data have been recorded onto magnetic tape. Every new successful shot of antiprotons and protons contributed slightly to the data sample and if the predictions were right, and the fluctuations of statistics were not too unkind, the first W boson was expected soon. The early part of the run included many machine development periods of one day, interspersed between the data-taking periods. At the time this was very unpopular with the experimenters although these development periods did produce very significant long-term improvements in beam intensity.

As the days went on the experiments continued to run reliably and the 'express line' tapes were analysed for all W boson candidates within a day or two of the data taking. After some very successful machine development sessions the beam intensities were further increased to a new peak value. The number of collisions was also increased by switching on the focussing magnetic quadrupoles on either side of the intersection region, which squeezed the beams to even smaller sizes at the centre of the detector. The accelerators and experiments continued to run successfully on a more regular basis, the beam intensities were now more consistent and the beam transfer procedures were becoming more reliable. People became more and more confident that the W boson would soon be found.

11.4 The first W candidates

Although the running and the computing seemed to be going well, there were no signs of any W bosons in October 1982. There were not even any events that were remotely like W bosons, or even the expected background to them. As the electroweak theory had been very successful, some people thought that the non-observation of the W boson would be even more interesting than its discovery. This group did not include many of the people working on the collider experiments, who certainly hoped that the W boson would appear soon. The high-precision graphics displays were being used day and night to check carefully for any possible W candidates. The scanning was usually done by a pair of physicists and as there were no really exciting W candidates much of the scanners time was spent on the events with jets. These events were very dramatic and were now being recorded with even higher energies than the previous year, but were unlikely to yield evidence for the W boson. Although these had been a particle physics highlight in the previous summer, they did not seem so interesting in the middle of a dark November night.

Then at last in early November things changed. Suddenly, during one of these routine sessions, a W candidate was found. This created great excitement and messages to the experimental control room, and soon there were tens of physicists, eager to look at every feature of the event, in the scanning area. The versatile displays are able to present numerically and visually all the raw information recorded by the detector and also the results of the full event reconstruction. An experienced physicist can then check very rapidly for any problems caused by faults in the experimental detector or reconstruction programs. These visual displays make the best use of the experimenter's skill in relating many different pieces of information and searching for possible explanations of an observed effect.

What aspects of this event caused so much excitement? It had many recorded hits in the central track chamber and these had been reconstructed to show the paths of many particles emerging from a common vertex, where the collision had occurred. Of course, the beam pipe goes right through the centre of the detector and so the beam particles and the actual interaction point can never be 'seen'. One of the tracks was particularly straight, so it clearly had a very high momentum and this was confirmed by the reconstructed momentum of over 40 GeV, which was printed together with the track on the screen. Sometimes the high momentum is fictitious and caused by some error in the choice of hits along the track and so this was checked carefully. Careful study showed that the track was clearly defined and the magnified display of points along the track indicated that it had been correctly reconstructed. At the end of each track a dotted line indicated the point where the particle would strike the electromagnetic calorimeter. The energies are displayed with the same shape as the calorimeter cell, but with a size which depends on the deposited energy, so that the energetic cells are prominent on the display. Just where the high-momentum track entered the electromagnetic calorimeter, a similarly large energy had been deposited. It was the first really energetic track that had appeared in all the events selected, and it was at first sight quite consistent with the expected fast electron from a W boson decay.

Then the detailed deposition of energy in the calorimeter was studied more carefully. An electron always gives up its energy in the first three depth samplings of the electromagnetic calorimeter and not into the fourth sampling nor the hadron calorimeter. Closer study showed that the hadron calorimeter cell in the region of interest had received small but significant energy, and so this particle could not be reliably classified as a single electron. It was the closest event yet recorded, it was the first serious background event that had been found, but it was not the first observation of a W boson. As we could not be certain it was an electron, the W candidate had to be rejected and the search continued. Many of us were in the scanning area for very long hours during this phase of the experiment, trying to make it easier and faster to check all features of an event. The first W candidate, which I have just described, was discovered by a colleague from Rome and myself on a routine scanning shift. After the near miss it was hard to get back to recording the details of more typical events, but that is what we did after the excitement had died down.

The scanning room is close to the main cafeteria, which remains open until two thirty in the morning during weekdays. The latest news from the experiment was frequently discussed over a late beer in the cafeteria but if you missed that deadline, then a coffee machine was the only convenient alternative. Many people, especially those visiting CERN for short periods, would be working very long hours to get their projects completed and as sleep became a rare commodity the flight back from Geneva always offered a very welcome rest. The run and the associated computing and scanning continued to go well, but no new W candidates were found.

Two days later a second W candidate was found in another routine scanning session, and this generated great excitement and very detailed studies were made of all aspects of this new event. An energetic track hit the electromagnetic calorimeter exactly where an enormous energy was deposited and immediately the experts on the calorimeters and the central detector were called in to check the technical details of the event. This time the energy was only deposited in the first three samplings of the electromagnetic calorimeter and absolutely none was recorded in the associated hadron calorimeter cell. The central detector had recorded other charged particles but these all had low transverse momentum, except for the electron candidate, which had a transverse momentum of 40 GeV! The calorimeters record the energy flow of both charged and neutral particles and the only large transverse energy was deposited in the cell struck by the electron candidate. This really looked like an electron as its momentum was the same as the energy deposited in the electromagnetic cell. As this was the first serious candidate for a W boson decay to an electron, you can imagine the excitement. A candidate event for the decay of a W boson to an electron is shown in figure 11.1. Copies of the event, as viewed from every angle, were made and they were distributed around the collaboration very rapidly. The phone was ringing constantly with people interested to hear the latest news and the whole of the collaboration seemed to pass by the scanning area on that day. Each person who studied the event tried to spot a problem that everyone else had missed, but none was spotted after a close inspection by many experts, and so this event was the first excellent candidate for a W boson decaying to an electron.

Why was there any doubt at all? The W^- boson was expected to decay to an energetic electron and an antineutrino. The antineutrino escapes from the detector without interacting and so its presence can only be deduced indirectly by using momentum conservation and assuming nothing else has escaped detection. The UA1 experiment was the first to try to 'measure' a neutrino by such an approach and no one was sure that it would really work. In this event, when all the calorimeter cells were added together to measure the energy flow, there was a large 'missing' transverse momentum opposite to the direction of the electron. This was consistent with the presence of an energetic neutrino being emitted back to back with the electron.

Any experimenter is keen to extract as much information as possible from the data recorded by a detector, but the neutrino measurement at first sight seems to be very difficult. Imagine that the large energy deposition, which we have identified with an electron, was due to a passing cosmic ray and was nothing to do with the event. Then, when we applied momentum conservation, we would incorrectly deduce the presence of a neutrino. Suppose part of the detector was not working correctly, then again we could easily make a serious mistake in the measurement of our 'neutrino'. The detector is virtually closed around the collision point, but there are a few gaps even in the calorimeters and if energetic particles choose to travel into these regions, then we will deduce an incorrect missing energy. For these reasons it was never likely that a single event could be conclusive proof of the existence of a W boson. However, if several were detected, many of these possible backgrounds can be eliminated by studying the distribution of the activity around the detector.

However, there were already some very encouraging signs from this first W boson candidate. The 'neutrino' was not opposite a known gap in the detector so it was unlikely that energy had leaked out. The comprehensive nature of the detector now became a big advantage, as the energetic electron was detected independently by

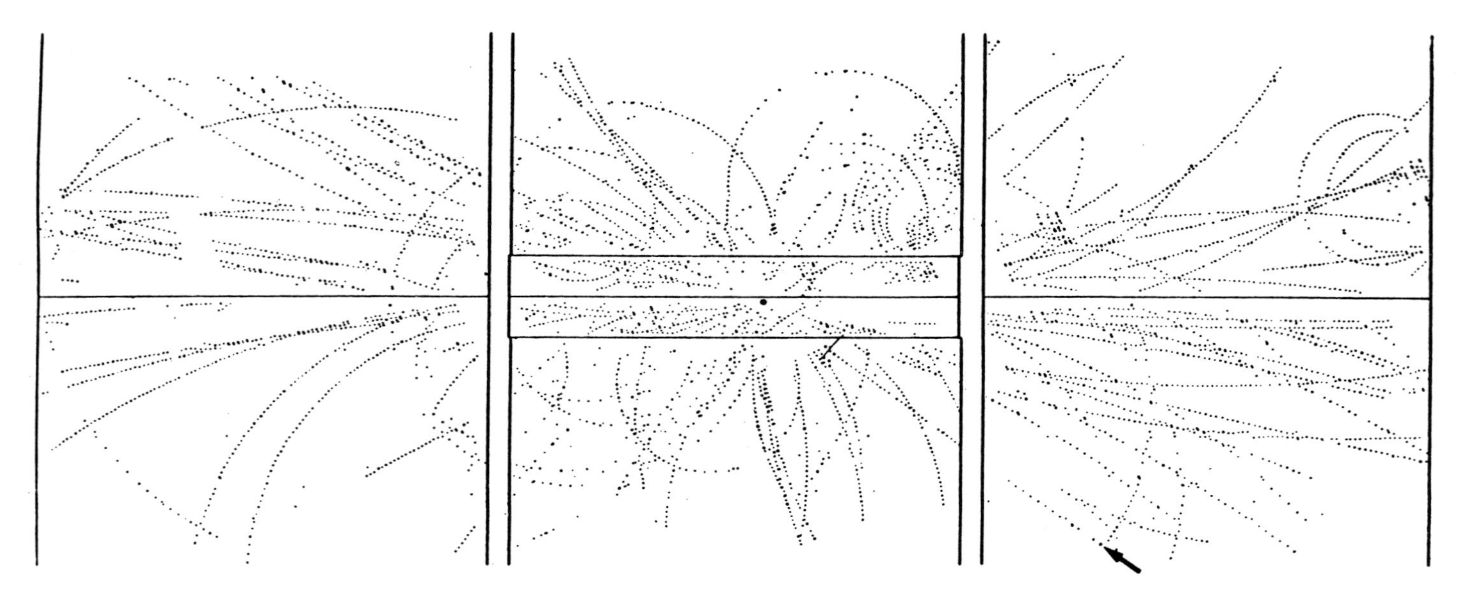

Figure 11.1. The charged particles detected by the UA1 central detector in one of the earliest examples of W boson production in antiproton–proton collisions at 540 GeV. The arrowed track is the straight, high-momentum electron resulting from the decay of the W boson, and the other particles are all fragments of the violent collision. The central detector needs to be very carefully calibrated and the analysis programmes have to be very sophisticated to deal with such a complex collision.

the central detector and the calorimeter, and so the chances of a strange detector problem causing both of these was vanishingly small. In addition the charged particle clearly pointed exactly at the collision point so it was almost certainly associated with the antiproton–proton collision. The observations seemed correct, the energy was associated with the event and so there were two simple explanations. Either momentum was not conserved or one or more energetic particles had escaped detection. The W boson was predicted to decay to an energetic electron and antineutrino and so this seemed the most natural explanation for the event. However, as it was only a single event, it was not conclusive.

This event had really lifted the morale of the collaboration, but more data were needed to establish whether the W boson had indeed been discovered, and so the experiment continued in an even more determined mood. The W boson decays to a muon and an antineutrino would also have been very useful evidence but these events were not being selected onto the express line tapes at this time. In a small fraction of the events there was evidence for the detection of two independent muons and only these were selected for express line analysis in this run. This meant that any supporting evidence from the independent muon decay would have to wait until the end of the run for the normal data tapes to be processed.

11.5 Can more be found?

The W candidate was a major boost to the spirits of the collaboration and there was still over a month of the experimental run to go before the collider was scheduled to stop on 6 December 1982. However, in such a large project there are so many uncertainties and one major problem with the detector or the accelerator could bring the whole operation to a sudden halt. If all went well, it looked as if the number of events could be increased by a factor of five before the end of the run. Five more W candidates would be very welcome, but with such small numbers almost anything can happen. It was obviously very important to optimise the data taking and increase the number of W candidates in the final month of the run.

There are two separate ways to optimise the operational efficiency of any detector. One is to minimise the changes in trigger conditions and any collision time spent on taking data in non-standard test conditions. The second is to keep the 'dead time' of the detector as small as possible by only recording a small fraction of the collisions onto magnetic tape. If the selected fraction of data is increased, the data acquisition computers become more active and so there is an increased chance of the system being 'dead' or busy when a really exciting event is produced. From the earliest days it had been the policy of the UA1 collaboration to minimise the changes of trigger conditions and to keep the 'dead time' below 10%.

These were sensible policies but there are times when exceptions need to be made. Remember, only the triggered events are recorded so that any collision that fails this selection is lost for ever! As there are so many interesting processes to be studied, there are many ways that the trigger can be extended to include quite different types of collisions. Also the analysis of the final data crucially depends on the triggering electronics for the muon and calorimeter events being well understood. Often these can only be thoroughly checked by recording a small number of events under special trigger conditions. These runs are essential, but occupy valuable beam time and there is always the risk of missing a rare W or Z boson event. The trigger conditions were a regular source of heated argument at the daily meetings. When the case was good enough, special test runs with changed trigger conditions were made. The policy of stability of triggers and a dead time of less than 10% was an important factor in maintaining the overall data taking of the UA1 experiment at an efficiency of around 90%.

Even in normal data taking there is always a difficult compromise to make in the events that are selected by the trigger. Consider the trigger for electron candidates, where a minimum transverse energy has to be detected before the event is recorded. If this threshold is too low, too many events are recorded and the risk of missing a really interesting event is increased. This problem can be removed by raising this threshold, but then it will be more difficult to interpret this highly selective data. If the beam intensity increases then the thresholds have to be raised to keep the dead time of the experiment small.

The muon trigger cannot be adjusted in such a simple way by changing a single threshold. The electronics, which identify muon 'tracks' which point towards interaction, select muons with a wide range of momenta. When the beam intensity was increased this created a real problem because the rate of muon triggers became too high. The main contribution to the muon triggers came from the chambers near the beams and so these were excluded from consideration. Events with muons were recorded only if there was evidence for muons in the chambers far away from the beam, in order to maintain the trigger rate at a reasonable level.

The daily schedule continued and the shifts became more routine, but the experimenters were beginning to tire as it had been a long period of preparation and data taking. The pressures at run time are very great because everyone knows that any unsolved problems will result in some useful collisions being wasted. Every few days an interesting new event would emerge, sometimes a new electron candidate for W decay and at other times an event with particularly energetic jets. The W candidates were accumulated and were being studied in great detail and the preliminary estimates of possible backgrounds to this signal were made. A further W candidate event is shown in figure 11.2.

The scheduled end of run date of 6 December 1982 approached and most of us had very mixed feelings about it. It had been a long tiring run and many people were looking forward to a break from the relentless daily pressure, but on the other hand the detector and the accelerators were working really well. A few extra days at the end of a run are often worth a few weeks at the start, and so the request for an extension of two weeks of running was made. This seemed a very reasonable request, as the experiments were on the brink of important discoveries. However, there were other vital jobs that needed to be completed before Christmas, so after due consideration the request was refused and the run was terminated as scheduled.

The accelerators at CERN are switched off for January and February each winter, to reduce the power consumption in these months when electricity demand is high. This meant that the whole resources of the UA1 and UA2 experiments could switch to analysing and interpreting the data that had recently been recorded. The detectors had been running from September to December and had each been exposed to 1000 million collisions during this running period. In the UA1 experiment, one million of these collisions had been considered interesting enough to record on magnetic tape and, of these, around 150000 of the most energetic events had been fully reconstructed on the central CERN computers. Thousands of the most interesting events had already been scanned in the search for a few W boson candidates. The total numbers of recorded events is very large and so it takes a great deal of care and checking to ensure that the few precious W events are not lost by one of the many selections that have to be made. By the end of the run in early December we had collected a total of six candidates for the decays of W bosons. These included examples of both $W^- \to e^- \bar{\nu}_e$ and $W^+ \to e^+ \nu_e$.

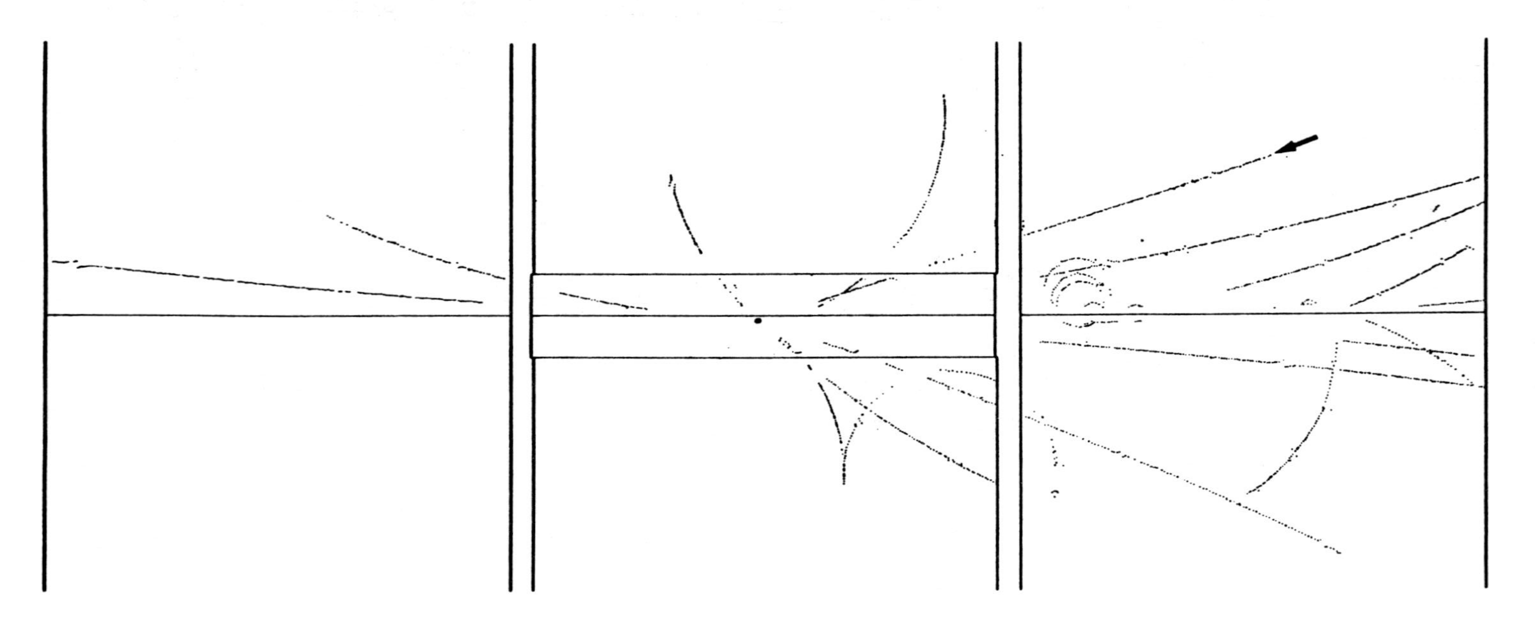

Figure 11.2. Another early example of W boson production but in this case the number of extra tracks is very small. The energetic decay electron from the W boson is arrowed. Each dot along the track corresponds to an electrical signal on a different wire.

11.6 Is there evidence for the W boson?

Although the run was now over, the pressure was still intense as the priority switched to a rapid completion of the analysis of the recorded data. Preparations were being made to start the analysis of the muon data, but the only chance of an early discovery rested on the decay of the W^- to an electron and antineutrino or the W^+ to positron and neutrino. The full sample of 28 000 events, with a localised transverse energy deposit in the electromagnetic calorimeters of more than 15 GeV, was rapidly reconstructed at the CERN computer centre in a systematic search for the W boson. In order to extract the rare electron candidates from the background of pions, several selections needed to be made. When an energetic charged particle with transverse momentum above 7 GeV was required to point to the largest calorimeter energy deposit, this left only 1106 events. In order to reduce the risk of other particles faking the energetic electron, the track and cell energy were required to be isolated from other energetic activity and finally the track was required to agree very precisely in position with the energy deposition. These selections yielded just 167 events, which were further reduced to 39 events, when the track momentum was checked for consistency with the electromagnetic energy and the associated hadronic energy was required to be small.

The surviving events were individually examined by physicists, on the interactive graphics terminals, in order to verify that these were genuine antiproton–proton collisions and that no reconstruction errors had occurred. The events divided clearly into two classes, 6 events with no jet activity and the other 33 with one or more jets of localised calorimeter energy. The events with jet activity had no missing transverse energy, but the 6 W candidates showed evidence for a missing transverse energy of the same magnitude as the electron. This is illustrated on figure 11.3, which shows that in these events the missing transverse energy is back to back with the electron in the transverse plane.

The W candidate events were then subjected to the most critical scrutiny. The most difficult people to convince are the experimenters themselves, as they have constructed the detector and know its limitations very well. These candidate events were clearly unusual but could they be due to something other than the W boson. As the W boson decay to an electron also included a neutrino, the unambiguous identification of W decay was bound to be difficult with just a few events. All the other possible explanations for the W candidates had to be considered and the probabilities for each of these separately

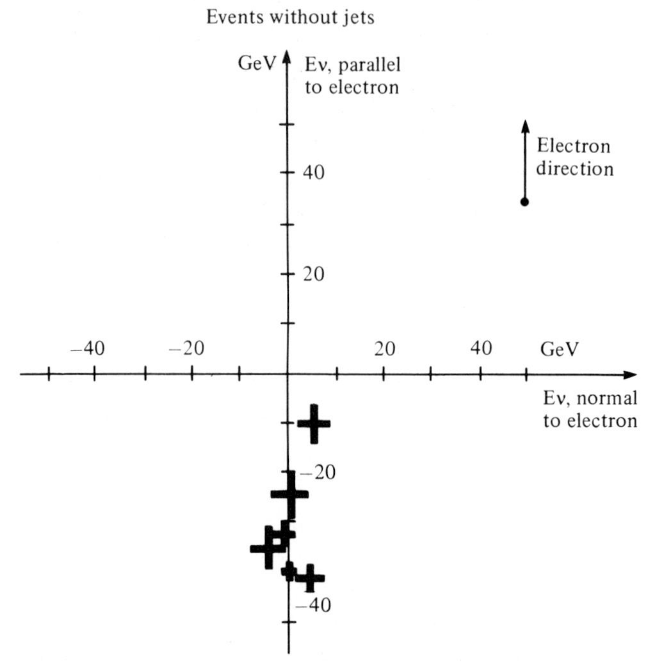

Figure 11.3. A study of the missing transverse energy of the six candidates for W boson production. The missing transverse energy is separated into components parallel and perpendicular to the electron in the transverse plane. As expected for W boson production the only significant missing transverse energy is parallel and opposite to the electron. This is consistent with the decay of a W boson to an electron and an energetic neutrino (from G. Arnison *et al.*, *Phys. Lett.* **122B** (1983) 103).

calculated. These other explanations are called backgrounds as they could in principle be responsible for an incorrect interpretation of the W candidates.

The main background to electron identification is provided by charged pions, which are produced 1000 times more frequently than the electron in antiproton–proton collisions. An energetic pion usually deposits its energy in calorimeters in a different way to an electron, as it deposits some energy in the hadronic as well as the electromagnetic calorimeter. Careful studies of the calorimeter in a test beam showed that very occasionally the pion could behave like an electron. However, the characteristics of these few W candidates were in excellent agreement with the expected electron behaviour, so it seemed unlikely that they were misidentified pions. How could we be sure that these W candidates were electrons? The discovery of jets of particles now became a useful tool in the study of the W boson, as practically all of the events which contained energetic particles also included jets. Although an energetic pion could occasionally masquerade as an electron, most of these energetic pions were produced in jets. The probability of observing an energetic pion in an event without jets, which also faked an electron, was extremely low. It is also possible for both a charged and neutral pion to strike the same cell of the calorimeter and fake an energetic electron but this background was also shown to be very small.

We have already seen that many systematic checks are necessary before the 'missing energy' can be reliably attributed to a neutrino or other non-interacting neutral particle. However, W candidates had already been studied very carefully to ensure that there were no inconsistencies in the events or in their reconstruction. The missing energy was soon validated and so each event contained clear evidence for an energetic neutrino as well as an energetic electron.

These events were beginning to look very convincing as W boson candidates but the intensive studies continued; how many other events had such large missing transverse energy? All the express line data, with evidence for at least one high transverse momentum track, were reanalysed and a search was made for events with large missing transverse energy. Would this yield a large background sample to the small number of W candidates? This independent search was performed in the few weeks around Christmas in 1982 and the results were dramatic! Only 18 events were found with a validated missing transverse energy exceeding 15 GeV, and of these only eight events had no jets and six out of the eight events were in our previous sample of W candidates! The additional two events had failed our test on energy matching between the charged track and the calorimeter energy deposit and were potentially candidates for the more complicated W decay to the tau lepton.

The same physical process that generated the high-transverse-momentum electron also produced a high-transverse-momentum neutrino. This missing energy selection also provided another result in the verification of the events. In the missing energy search no requirement at all had been made on the 'electron-like' nature of the calorimeter energy. When the eight events without any jets were examined, their calorimeter deposits were exactly as expected for electrons, even though no such requirement had been made in the selection! These really looked like genuine electrons and serious candidates for the first recorded examples of the production of the W boson. Over this same period the UA2 collaboration were also rapidly analysing and checking their data. It was known that they had also seen interesting candidates but there were no formal presentations by either collaboration until January 1983.

11.7 The W boson is discovered

A workshop on antiproton–proton collider physics was held at Rome University from 12–14 January 1983. This was attended by many of the experimenters who had participated in the long periods of detector preparation and data taking at the end of 1982 and it was a very productive meeting. The seminars were well attended and held in a large lecture theatre in the Physics Department of the university. There were a number of talks on both theoretical and experimental topics but all were concerned with energetic antiproton–proton collisions. Many of the talks reported on the recent antiproton–proton measurements made at CERN and how these compared with theoretical expectations.

The most exciting session included the presentations of Carlo Rubbia and Pierre Darriulat, who described the status of the searches made by the UA1 and UA2 experiments for the W and Z bosons. There had been intensive work by both collaborations to complete these status reports so soon after the data-taking period. The atmosphere was electric, neither collaboration were sure of the conclusions of the other, and so the audience listened intently. The UA1 talk was presented at breathtaking speed by Carlo Rubbia who managed to compress several hours of material into one hour. The whole UA1 experiment was briefly reviewed, from the properties of each part of the detector, to the final selection of W candidates. One thing was evident – there was not a single candidate for a Z^0 boson but there were six golden W boson candidates, of which five were quite clearly inconsistent with other known processes. Finally all the possible background explanations were discussed, together with their likelihood. Even the most pessimisitic estimate for the background suggested that it was less than one event! The conclusion was cautious, these events were perfectly consistent with the decay of the W boson. All known sources of backgrounds could not explain them but had something been missed?

The UA2 talk was also very comprehensive but was presented at a more leisurely pace. Would this independent experiment reach quite different conclusions? The early part of the talk was concentrated on the search for the Z^0 boson; had it been discovered by UA2? After the selections had been described, the UA2 conclusion on the Z^0 boson was that no candidate event had yet been isolated. The W boson search was then described in detail and four excellent W candidates were presented. The background calculations were made and again the conclusion was that the observed events were consistent with the predicted decay of the W boson and inconsistent with any known background. The conclusion from the UA2 collaboration was also very cautious and it was stressed that more work was needed to check for other possible backgrounds.

At lunchtime after this session, the conversation was dominated by the W boson results. Ten candidates for decays of W^- to electron–antineutrino and W^+ to positron–neutrino had been presented, with a calculated background of less than one event. What else could these events be – surely the W boson had been discovered? Considering the complexity of the detectors it was a tremendous achievement to have fully analysed the W candidates in such a short time, as the Rome meeting was barely one month after the end of the run! This was emphasised in the final session of the meeting by Leon Lederman, the director of the Fermi National Accelerator Laboratory in the USA, who said that he was impressed by 'the speed at which the data were analysed and physics achieved, out of detectors of unprecedented sophistication, viewing collisions of novel complexity'.

When the collaborations returned to CERN, further details were checked and cross-checked and no further problems were identified. The mass of the W boson could even be estimated by combining the electron and neutrino measurements. The errors were large, as there were only a few events, but both collaborations were able to quote the mass of the W boson as being close to 80 GeV. The results from the two experiments were presented at seminars in the CERN auditorium on the 20 and 21 January 1983. The UA1 talk was presented by Carlo Rubbia and the UA2 talk by Luigi di Lella. These two outstanding talks were attended by far more than the 400 that can be seated in the auditorium. The room was packed, with many standing, and the atmosphere was tense and exciting. The presentations were still tentative and qualified, could there really be another explanation for the candidate events?

The UA1 collaboration, over the weekend of 22–23 January, began to write a scientific paper on the events. Carlo Rubbia was quoted as saying 'they look like Ws, they feel like Ws, they smell like Ws, they must be Ws'. All the experimental data and the background calculations had to be collected, checked and described. Anyone who had been involved in any aspect of the analysis was frantically trying to get their part of the work completed as rapidly as possible. There were numerous meetings at all hours where all the remaining tasks and difficulties were discussed. When anything is done at this speed, it is very difficult for everyone to be fully consulted but the text and the figures for the publication were rapidly assembled and then discussed in further meetings. The people most involved in the

Figure 11.4. A picture taken at the press conference held on 25 January 1983 at CERN to announce the discovery of the W boson. From left to right the participants are Carlo Rubbia (spokesman of the UA1 experiment), Simon van der Meer (responsible for the antiproton accumulator), Herwig Schopper (director general of CERN), Erwin Gabathuler (research director of CERN) and Pierre Darriulat (spokesman of UA2 experiment). The diagram indicates the locations of the UA1 and UA2 experiments around the circumference of the super proton synchrotron. Photograph courtesy of CERN.

preparation of the paper had been working so hard on the topic that some of the meetings were very tense. The analysis of the data had now been available for several weeks, so there had been plenty of time to discover serious errors if they existed. Finally the draft publication was produced and discussed with the collaboration at a hastily arranged meeting where the publication was agreed. On the 25 January a press conference was called at CERN, to announce formally the discovery of the W boson (figure 11.4). At this stage the UA2 team preferred to reserve judgement, but further analysis convinced them also. Both teams were able to quote masses for the W boson which were consistent with the predictions of the electroweak theory. The press release read as follows:

Europe forges ahead in particle physics

Results long awaited by physics have been obtained at CERN, Geneva. Experiments conducted here by international teams of scientists using matter/antimatter collisions begin to reveal the expected signature of a long sought particle of matter: the W intermediate vector boson. The importance of the W is that it gives validity to the electroweak theory unifying the description of 2 of the 4 basic forces of nature.

CERN – the European Laboratory for Particle Physics – has set up two large underground experimental devices on its unique SPS collider. This ring-like machine 2.2 kilometres in diameter straddles the Swiss–French border near Geneva. There, particles of matter called protons are smashed against their antimatter counterparts – the antiprotons – obtained from a source making use of a technical CERN invention: the 'Stochastic Cooling' process. This allows large numbers of collisions to take place.

The collisions take place in underground caverns where elaborate detector systems record the features of extremely rare and evanescent building blocks of matter such as the bosons. These experiments are carried out by two large experimental teams, UA1 and UA2, gathering a total of 180 Physicists from 8 European countries and the United States. The teams have presented the following results at CERN.

UA1 has singled out 5 events in a total of a thousand million collisions, revealing the expected signature of the charged W boson. The results also give a determination of the W mass of the order of 80 GeV (80 times the mass of the familiar proton), which is in agreement with the predictions of the electroweak theory.

The UA2 team has observed 4 events from the same number of collisions that are consistent with a W signature, but more work is needed to confirm this preliminary UA2 result.

A forthcoming operating period of the collider should confirm the results of the observed events and provide additional information through other features of the W decay. It is also expected to observe the neutral intermediate boson, called Z, which can be produced less frequently than its charged W partner.

The calculated gamble of the CERN directorate in backing the collider programme had paid off. The diverted resources and manpower had been used to produce a major international success. It had not been an easy decision as any major faults in the accelerators or detectors could have prevented the successful completion of the project. However, each stage had been very well planned and coordinated and each major uncertainty had been carefully evaluated. The enthusiastic efforts of an enormous number of people had been harnessed effectively. The successful discovery of the W boson was a reward for team work on a scale unprecedented in scientific experiments.

Chapter Twelve

The search for the Z boson

12.1 The winter shutdown

After all the excitement of the January meeting in Rome and the publication of the discovery of the W boson, it was hard to get back to routine analysis of the rest of the experimental data. There was still lots of work to do and many events that needed analysing. For these analyses it was necessary to use all the tapes of data, which contained over one million triggers, selected from over one billion collisions. The full reconstruction of an event takes up to 20 seconds on the largest computer and so the standard reconstruction of all the events was clearly excluded if rapid results were to be obtained. The analysis of the events containing a muon candidate was started in early January together with analyses of the properties of jets and many other interesting types of events.

After the winter shutdown of the CERN accelerators, which lasted until the end of February, the first month of super proton synchrotron (SPS) running was scheduled for normal fixed-target experiments. This seemed quite reasonable as a way of sharing the accelerator's resources, but it meant that the operation of the accelerator had to be changed yet again. The next collider run was scheduled to start on 12 April 1983 and was due to last until July. The immediate target of the run was very clear. It was hoped to accumulate many more examples of W boson decays and to find the Z boson.

12.2 What do we hope to observe?

Over one billion antiproton–proton collisions had been detected in the experimental run of 1982. Not a single candidate for the production of a Z^0 boson had been found in either of the two experiments. This was disappointing, but not altogether surprising, as only nine W boson events had been produced, five from UA1 and four from UA2. The theoretical predictions were quite clear, the Z^0 boson should be produced in an observable way, only one-tenth as often as the W boson. It seems that nature chose to reveal the W first and reserve the Z for another run. This certainly extended the period of excitement and allowed the W discovery to be carefully studied in isolation.

If the beam intensity could be increased and maintained in a future run, then a few Z^0 bosons should also be produced, but would this be enough to prove the existence of the Z^0 boson? From the earliest proposal it had always been expected that the background to the Z^0 would be much lower than for the W boson. The Z^0 boson certainly had one major advantage over the W boson, as it was expected to decay into two charged particles and not the elusive neutrino. The signal for a Z^0 decay is truly dramatic as the massive particle converts virtually all its mass energy into the motion energy of an electron–positron or muon–antimuon pair, which are emitted back to back if the Z^0 is at rest. In both cases the event should contain two energetic charged particles, and in the electron decay these will deposit very high energies where they enter the electromagnetic calorimeter. In the muon decay the particles will deposit a small amount of energy in the calorimeters and then produce hits in the muon chambers, which are outside the calorimeters. The mass of the Z^0 boson could be calculated from the energy and momentum of its decay products and was expected to exceed 90 GeV!

12.3 The 1983 run starts

As the run approached the emphasis was switched from analysis back to preparing the experimental detector for another major run. Gradually the meetings were transferred from the main CERN site, back to the small conference room near the experimental area. The meeting topics changed from the latest news on the physics analysis to planning of the installation of the detector. One brief look at the complexities of the electronics and associated computers would emphasise that it is not straightforward to restart the experimental programme of such a large detector. The system is continually being developed and improved, so there are always new procedures to be introduced at the start of a new run. The experimentalists had now been involved for over 6 months in preparations for the previous run, the run itself and a highly intensive period of analysis. They had been working under great pressure continuously since the previous summer. Some people viewed another 3 months of running with trepidation, but all were extremely keen to discover the Z^0 boson.

There was a lot of discussion, before the new run, of the changes that had been made to improve the beam intensity of the antiprotons. During each run some periods are used to investigate and develop new techniques for optimising the beam transfers and if these do not result in any improvement in beam intensity, these periods are very unpopular with the experimenters. However, if successful this investment of running time is repaid with interest, with improved beam intensities for the rest of the run. It is only during the experimental run that a reasonable prediction can be made of the number of antiproton–proton collisions that will be observed. The importance of stable and intense antiproton and proton beams cannot be overemphasised.

At the start of the 1983 run, even the peak beam intensity of the previous run could not be achieved. A series of different problems meant that very little useful new data were recorded, and we began to wonder why the whole operation had ever been shutdown for the winter! It would have been much simpler to carry on running successfully while everything was working well. Gradually things began to improve and the data tapes were spinning nearly as fast as before. The most exciting candidates, which included a small number of new W boson candidates, were selected in the same way as in the previous run and were being scanned just a day or two after they had been recorded. The W bosons were still sufficiently rare for them to excite some special interest, but the scanners quickly passed on to the next event in case it was even more exciting. It is said that 'yesterday's discovery is today's calibration and tomorrow's background'. This really seemed to apply to the W boson, where each new observation was really used as a check that the whole detector and analysis chain were functioning properly. However, by the end of the run the improved statistics of W boson events made it possible to test more detailed predictions of the electroweak theory and so the events were still carefully handled!

Steadily the beam intensities increased and the collider project began to run much more reliably, delivering a consistent daily supply of antiproton–proton collisions. The reliability of the antiproton accumulator was superb as it continued its daily cycle of collecting and cooling the antiprotons. The antiprotons were certainly well looked after in this device: in one period of over 30 days it operated without losing a single antiproton stack. The scanning and checking of events round the clock continued on two separate display screens. Several hundred events each day were being inspected and the scanning was only a few days behind the data taking. The selected express line events included events with energetic calorimeter energy deposits and those with evidence for two independent muons. The Z^0 was expected to decay equally into electron–positron and muon–antimuon and so it was hoped that some examples of both of these decays could be extracted before the end of the run.

12.4 A Z^0 candidate?

On 4 May 1983 the first Z^0 candidate was identified from the computer output of the analysis of a collision which occurred on 30 April. It was spotted, after midnight, by someone using the CERN computer from the Annecy laboratory. She immediately drove 40 kilometres to CERN to inspect the event on the interactive scanning

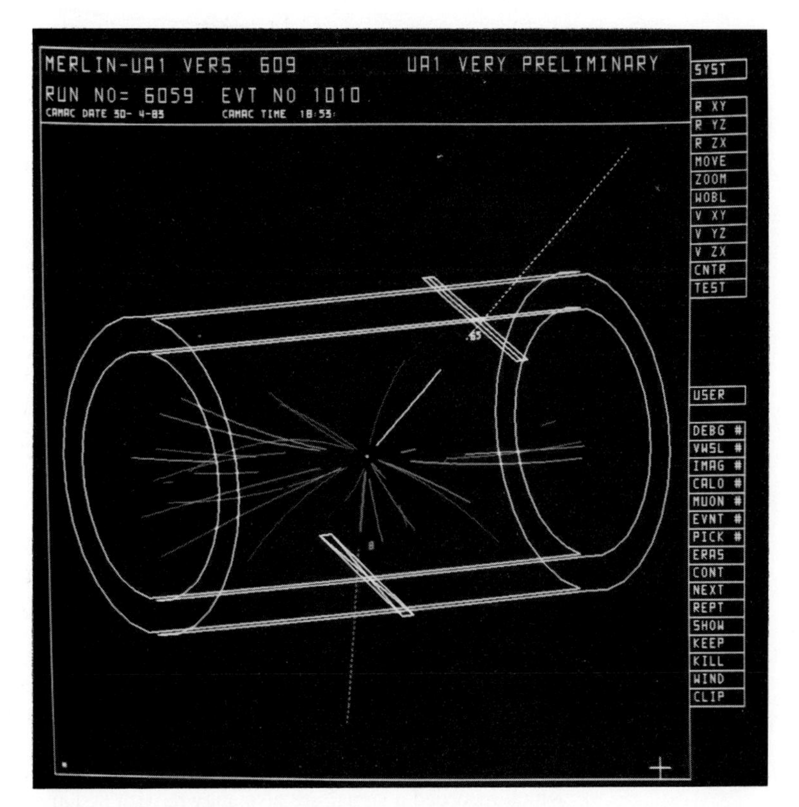

Figure 12.1. The charged particles recorded by the UA1 central detector in an early example of Z^0 boson production. The electron and positron decay products of the boson deposit large energies in the calorimeters (shown as rectangles). These are not back to back in this view because the quark and antiquark which produced the Z^0 boson did not have equal momentum. Photograph courtesy of CERN.

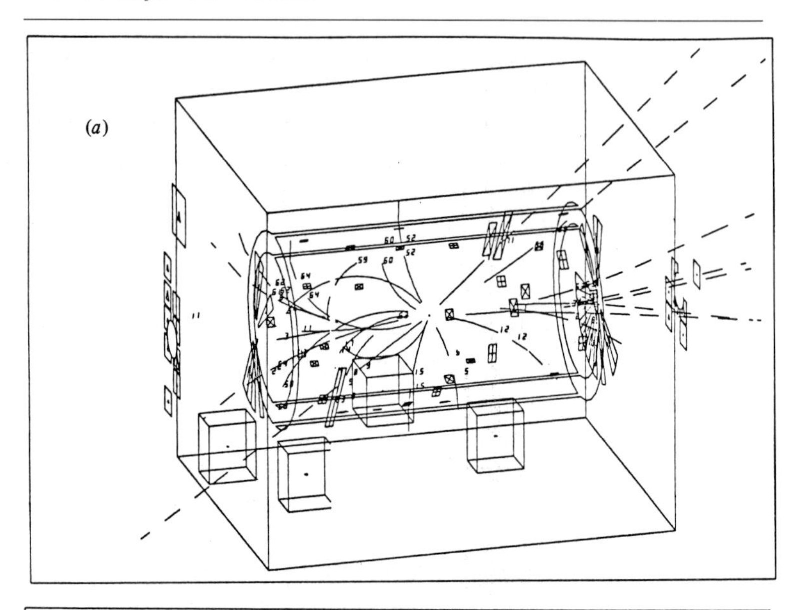

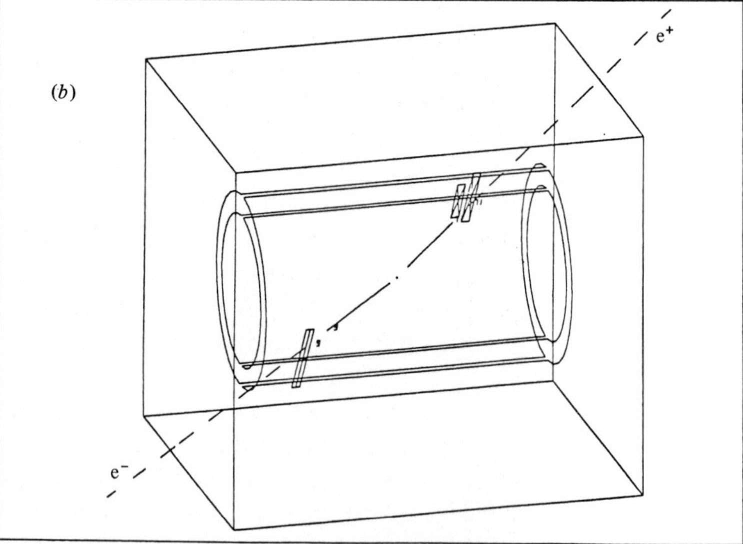

Figure 12.2. (*a*) A general view of the information recorded by the UA1 detector in another early example of Z^0 boson production. The solid lines represent the paths of charged particles traversing the central detector and the dotted lines indicate where these particles would have travelled through the calorimeter if they had not been absorbed. The calorimeter cells which recorded a signal are also shown, with a shape related to the geometrical shape of the cell. The rectangular shapes indicate energy deposited in the gondola and in this case the size of the displayed cell is related to the deposited energy. (*b*) The same Z^0 boson event where only tracks with transverse momentum above 1 GeV and cells with transverse energy above 1 GeV are displayed (from G. Arnison *et al.*, *Phys. Lett.* **126B** (1983) 398).

facility and was delighted to confirm it as a Z^0 candidate. The scanning area erupted with excitement – at last there was some indication that the Z^0 boson or heavy photon existed. As people crowded excitedly around the graphics monitor the dramatic features of the event became apparent. There were two straight tracks, at a large angle to the beam, in the central detector. Many other tracks were present, but none of these had a high transverse momentum (figure 12.1). These other tracks result from the violent interactions between the many constituents of the proton and antiproton. Usually only a single quark and antiquark annihilate to produce the massive boson, but all the other constituents are violently disturbed by the collision and these subsequently appear as fragments of the collision. The two tracks were 'back to back' when the collision was viewed from along the beam direction, and the two energetic particles had opposite charges. Just where the straight tracks entered the electromagnetic calorimeter, the gondolas recorded very large transverse energies, which were the only significant deposits of transverse energy recorded by the detector (figure 12.2). People asked anxiously about the hadron calorimeter cells behind the gondolas because if these had significant energy, then the Z^0 candidate would have to be rejected. The operator of the display is in a privileged position, at a time like this, right at the front of the crush. There are disadvantages though, as it is very difficult to satisfy all the various requests that are coming with increasing volume from all parts of the room! As the event was rotated to check the hadron calorimeter cells, first one then both tracks were checked. There was no energy behind the gondola cells, the tracks were clearly serious electron and positron candidates.

The gondola energy deposits were so large that it was very unlikely that any background process could produce this result. The quality of the two straight tracks was then carefully checked but both were well separated from neighbouring tracks and seemed to have been correctly reconstructed. There is an important difference between identifying an event as potentially interesting and the final calculation of its measured properties. In certain directions the final momentum of a fast track is sensitive to the detailed calibration of the central detector. The performance of this detector depends on the exact mixture of gases and other details of the operating conditions, which change gradually during a data-taking period. These changes are carefully monitored but there is always some delay before the final calibration is available. The accurate determination of the energy deposited in a gondola also requires careful analysis. The variation with time is less important here, but the final energy calculation is quite complicated and needs to make use of the exact entry point of the charged particle. Consequently the initial energy and momenta are useful guides but these need careful analysis to produce the best values.

We have seen that both gondola energy deposits were very high and that both tracks were straight. The transverse momentum of the positron, as measured in the central detector, agreed very well with that measured in the gondola which it entered. The transverse momentum of the gondola struck by the electron was also very large, 40 GeV with an error of a few GeV. However, the transverse momentum of this track was only 8 GeV! The track measurement is much less accurate than that from the calorimeter at high energies and we have seen that the final calibration for the central detector was not yet available. The background for producing the two massive energy deposits was so small that it was widely expected that the electron track would later be found to have a much higher momentum. Once the final values are available the track and calorimeter energies should of course agree for both the electron and the positron.

While these careful checks of energy and momentum measurements were being made, news quickly spread around the collaboration and CERN that the first Z^0 candidate had been detected. The news of this Z^0 candidate was greeted with spontaneous applause at a Science for Peace meeting, held in San Remo on 5 May, when announced by the director general of CERN, Herwig Schopper. At CERN the checks on this event continued, with all of the central detector and calorimeter specialists contributing to this work, and after many days of careful study the conclusion was unavoidable. The two large energy deposits in the gondolas were verified, together with the high momentum of the positron, as measured in the central detector. However, the electron had significantly lower momentum, as measured by the central detector, than that deposited in the nearby gondola. This seemed incredible; why did the first Z^0 candidate have to be so complicated?

One way to explain the anomalous track was that it was a charged track accompanied by an even more energetic neutral pion. This would not be visible in the track detector but would deposit energy with an electromagnetic profile in the calorimeter. There was a clear inconsistency because if this were true, then why did we not see many other less energetic events of a similar type? The event seemed to be a Z^0 as the background was negligible, but clearly the electron and positron were not carrying all the energy. There was another explanation for this event. When an electron or positron passes through matter it emits photons due to its electromagnetic interaction with atoms. These photons are emitted in a direction close to that of the electron but usually only carry away a small fraction of its energy. Any observed electron has to pass through the steel beam pipe and so it can emit a photon that travels along with it. If this photon took most of the original energy, then the electron plus the photon would produce the large energy in the calorimeter and the lower-energy electron would be recorded in the central detector. The difficulty with this explanation was that such an energetic photon should only be emitted once in every 100 Z^0 events. This event could be a Z^0 boson, but we were certainly hoping for a more straightforward event if any other candidates were found! The mass of the first Z^0 candidate was calculated to be 100 GeV using the approximate calorimeter energies.

12.5 Is it a second Z^0?

It had been a very long wait for the first Z^0 candidate but remarkably a few days later the first candidate for the muonic decay of the Z^0 boson was identified. It was found in an evening scanning session and led to a busy night. Again there were two very straight tracks amongst many other tracks of much lower transverse momentum. One of the fast tracks travelled upwards towards the top muon chambers and the other towards the muon chambers at the end of the detector. The scanners, plus several interested bystanders, carefully went through the routine checks which need to be made for each muon candidate. Each track passed through the electromagnetic and hadronic calorimeters and deposited a nominal amount of energy as expected for a muon. In each muon chamber, the hits are connected to show the direction of the particle that caused them. This muon track can then be compared with a dotted continuation of the central detector track to see if the two effects are related (figure 12.3). For both muons the match between the muon chamber and the central detector track was excellent so these appeared to be very promising muon candidates.

The detailed inspection of the charged tracks was encouraging as there was no evidence for any systematic problems, but on closer inspection the first problem appeared. The track pointing to the forward chamber was very well measured in the central detector and its charge and momentum could be measured directly. The other track was very straight, with an undetermined momentum, but it was so straight that even its charge could not be reliably measured. With a charge and momentum for only one of the tracks, it would not be possible to calculate the mass or charge of the parent particle!

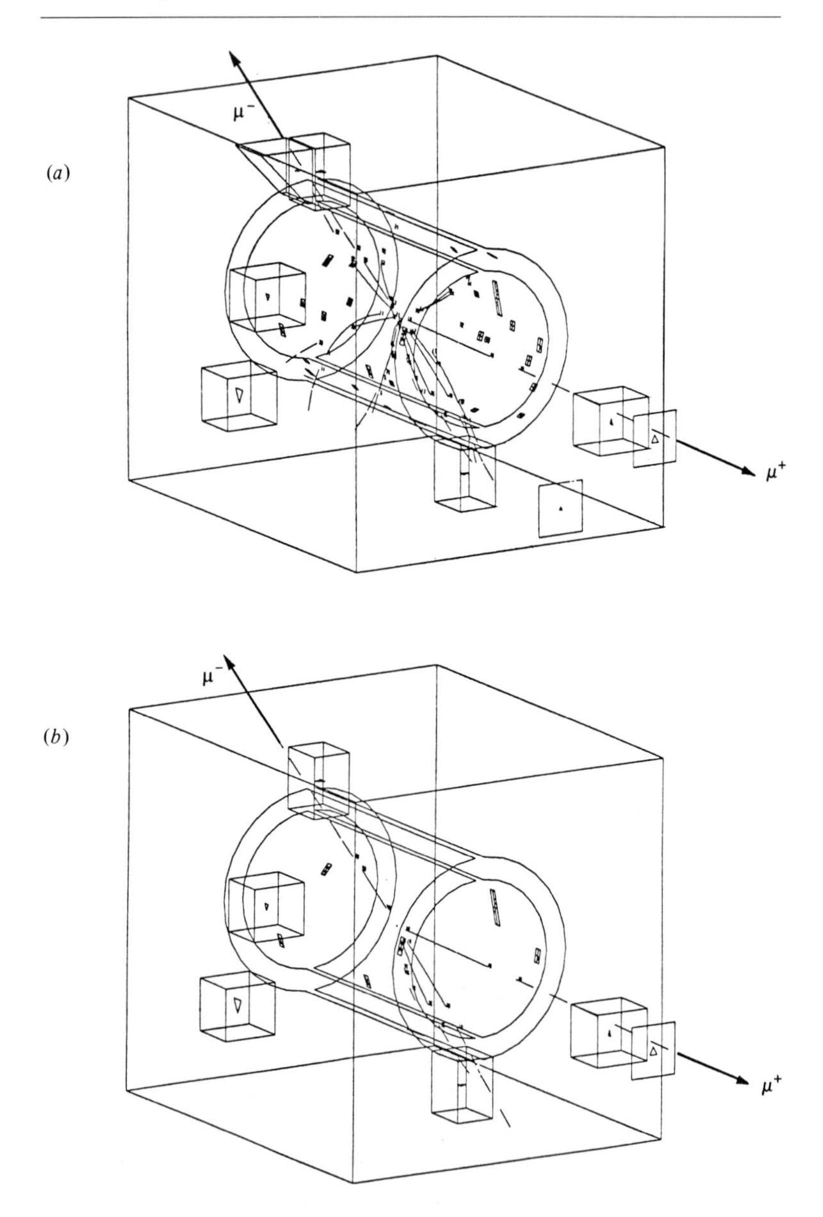

Figure 12.3. (*a*) An overall view of the information recorded by the UA1 detector for the first example of Z^0 boson production where the Z^0 decayed into a muon–antimuon pair. The symbols have the same meaning as in the previous figure but in this case the tracks detected by the muon chambers are shown as arrows. There is very good agreement between these tracks and the expected path of the muons from the central detector. (*b*) The same event with a 1 GeV threshold applied to the transverse energy of charged tracks and calorimeter cells. In this example the Z^0 boson was produced in association with a jet and so additional tracks pass this selection (from G. Arnison *et al.*, *Phys. Lett.* **126B** (1983) 398).

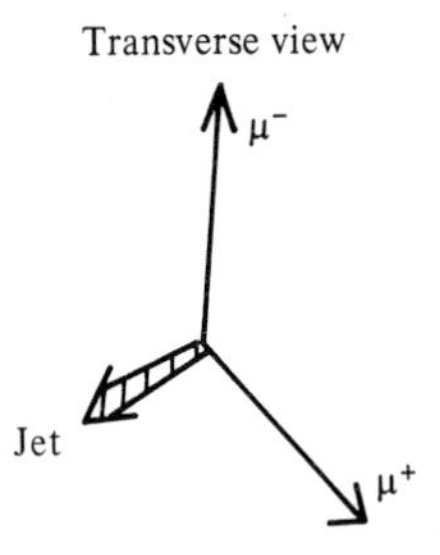

Figure 12.4. A transverse view of the event shown in figure 12.3 viewed along the beam direction. Usually the μ^- and μ^+ would be back to back in this view but the Z^0 boson is produced with some sideways motion which is balanced by a single jet of particles.

There was another surprise about this second Z^0 candidate – it had a clear jet of particles, balancing the sideways motion of the dimuon pair, as viewed along the beam direction (figure 12.4). This Z^0 candidate was proving even more complicated than the first. There was no reason why a Z^0 boson should not be produced with an extra jet, but it was certainly expected to be produced most frequently without one. How could we call it a Z^0, when it might not even be neutral and when we had no estimate of its mass?

We worked through the night on this event and by the following morning extensive progress had been made. There is a uniform magnetic field in the region of the central detector and also in the iron of the hadron calorimeter. All muons pass through this calorimeter and experience deflections which depend on their momentum and electric charge. By exploiting this information, the muon chamber measurements were used to measure the charges unambiguously and the two muons were found to be carrying opposite electric charges. This procedure also yielded an independent momentum measurement for the muons, which yielded a mass of the muon–antimuon system of 100 GeV, but how could these estimates of momentum be checked?

In Z^0 decay there is no neutrino involved and so momentum conservation can be used in a different way. The muons deposit only a nominal amount of energy in the calorimeters, so most of their momenta is not detected by the calorimeters. As the directions of both the muons were well measured (figure 12.4), the only unknown was the magnitude of the muon momentum of the poorly measured track. Consequently momentum balance could be used vertically and horizontally, using all the measured information, to yield two independent estimates of this muon momentum. These estimates were internally consistent and agreed with the calculation from the muon chambers. The mass of muon–antimuon system was calculated to be 100 GeV with a large error. This event appeared to be the first Z^0 boson ever observed to decay to a muon and an antimuon.

News of this second dramatic event also spread very fast around the collaboration. Both events were more complicated than we would have chosen, but they were quite different to anything that had been seen before. Because of the mounting excitement, a CERN seminar was arranged so that the UA1 results could be presented. This was another special occasion with standing room only in the vast auditorium. Carlo Rubbia gave an excellent presentation of the properties of these two events, leaving the audience to make up their own minds about whether these were indeed Z^0 bosons. At the time of this seminar the UA2 experiment had no evidence for any similar events.

12.6 The discovery of the Z^0 boson

The search continued for more Z^0 candidates in both experiments. Just a few days after the CERN seminar, two classic examples of a Z^0 boson decaying to electron and positron were identified in the UA1 experiment. Incredibly, after running for so many months these two events were recorded on tape within 15 minutes of each other! In these events the track and calorimeter measurements agreed well and the electron–positron features were very clear. Once these two simple Z^0 candidates had been observed it was clear that the Z^0 boson had been discovered. A single quark from the proton was annihilating with a single antiquark from the antiproton and producing a massive Z^0 boson which subsequently decayed. The dramatic localised energy depositions of the Z^0 boson events are shown in figure 12.5. Outstandingly high energies are deposited in the calorimeter by the electron and positron. The expected background was calculated and found to be insignificant. The detailed properties of the events were checked and cross-checked and during this time one more electron–positron decay of the Z^0 boson was found.

On 1 June 1983 CERN formally announced the discovery of the Z^0 boson. The press release began:

> CERN, the European Laboratory for Particle Physics, announces its second major discovery in the space of 4 months: that of the Z^0 intermediate boson, the particle which *carries the electro-weak force.*
>
> The Z^0 (zed zero) is a member of a trio predicted by an electro-weak theory to unify 2 of the 4 basic forces in Nature: the *electro*-magnetic and the *weak* nuclear forces.

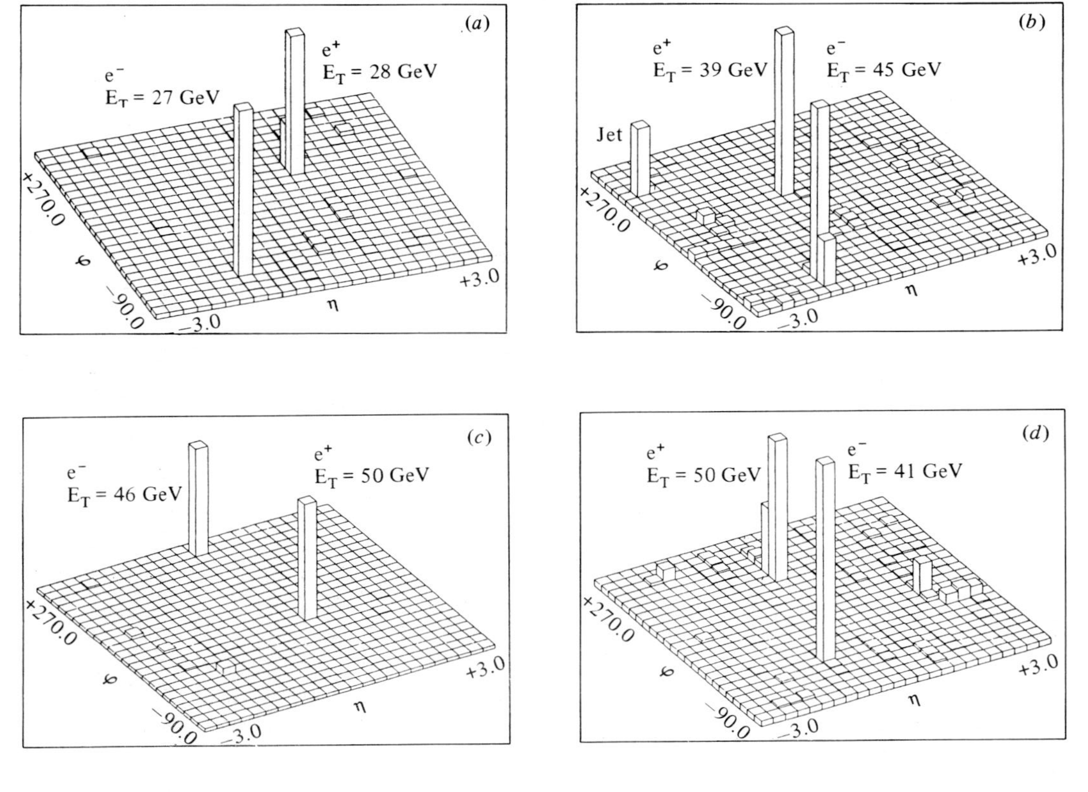

Figure 12.5. Illustrations of the localised and energetic calorimeter depositions in the UA1 detector in the first four examples (*a–d*) of Z^0 production, where the Z^0 decayed into an electron and a positron. The closed calorimeters are 'opened out' on to a plane for this display and the height of each cell is proportional to the transverse energy (E_T) deposited in it. The horizontal axis corresponds to cells along the detector (η) and the vertical axis to cells around the detector (ϕ). In three of the events there are no significant deposits of transverse energy away from the electron and positron. In one event an additional jet is produced (from G. Arnison *et al.*, *Phys. Lett.* **126B** (1983) 398).

In January scientists at CERN had announced the discovery of the 2 other bosons, the W^+ and W^-; 1983 now appears to be a vintage year of scientific progress made by CERN and European science.

Many people felt that a particle as elusive and as difficult to produce as the Z^0 could not be discovered so speedily. Its identification, following long periods of data analysis terminating over last week-end, is proof of the originality, selectivity and reliability of the equipment operated around the clock to translate into reality the brilliant ideas put forward by the boson hunters.

The UA1 collaboration hurriedly wrote a scientific paper on the discovery of the Z^0 boson which included four electron–positron decays and one muon–antimuon decay. This was submitted for publication on 6 June 1983 and quoted a mass for the Z^0 boson of 95.2 ± 2.5 GeV. It cautioned that as calibrations of the calorimeter were still in progress there may later be small energy scale shifts to both the W and Z boson masses.

The UA2 experiment detected several Z^0 candidates towards the end of the data-taking run. The distribution in time of these events had proved unfortunate for that collaboration but finally a similar number of Z^0 decays into electron–positron had been recorded by both detectors. Between them, the two experiments had collected nine Z^0 bosons with a mass centred on 93 GeV, which was in excellent agreement with the detailed predictions of the electroweak theory.

12.7 Summary

By the end of the run on 3 July most people who had been associated with the collider project were exhausted after a full year of experimental runs and hectic analysis. What a period it had been with the first ever observation of the W^-, W^+ and Z^0 bosons. Already several of the key predictions of the electroweak theory had been verified with the observation of the bosons at their predicted masses.

The enormous amount of data that had been recorded still contained a lot of new information which needed careful study. There were now over 50 W boson candidates waiting to be analysed. With this improved sample it might be possible to test the predicted 'left-handed' interaction of the W boson, which had not been possible up until now. It was time to search for the W decays into muons to see if the expected equal decay rate was observed. The elusive top quark had still not been found at electron–positron accelerators: could it be discovered at the collider? Even more exciting would be the discovery of a totally new class of events never detected before at lower energies.

It was time for a short break from the series of deadlines and frantic discusssions. Most of the collaboration disappeared from CERN for a few weeks for their vacations but a small group of people immediately started the analysis of the new data. The results of these studies are described in the next chapter but whatever else was found, 1983 was going to be a very hard year to follow.

Chapter Thirteen

The search continues

13.1 Introduction

Towards the end of the 1983 run an intensive programme of analysis was started on the vast number of data that had now been collected by the UA1 and UA2 collaborations. The running period of 3 months had allowed the study of 8000 million antiproton–proton collisions at the UA1 detector, of which 2.5 million were selected and recorded on several thousand magnetic tapes. Copies of the most dramatic events had been written to the express line tapes for rapid analysis and this selected sample of 115000 events was fully reconstructed within a week of being recorded. An international particle physics conference for over 700 delegates was to be held at Brighton, England, in late July 1983. The collection of the most interesting results from the collider run, for presentation at this conference, became a top priority for the two collaborations. Time was very short, but there was such wide interest in the experiments that it was important to include as many new results as possible.

This chapter contains a description of the new results on the W boson that were presented at Brighton less than a month after the end of the 1983 run. This is followed by an account of the search for the top quark, which continued until the summer of 1984. Finally, some very unusual events with large missing transverse energies are described, which are still being studied in 1985.

13.2 Extra information on the W boson

The W boson had been discovered in the data from the 1982 run but now there were many more data available from the UA1 and UA2 experiments. This larger sample of W bosons allowed further checks of the predictions of the electroweak theory. The electrons from the decay of the W boson were predicted to be emitted primarily in the general direction of the incident proton. This was an important prediction, as all other reactions which involve the weak interaction exhibit a similar property. The W boson is the force carrier for the weak interaction and so this asymmetry in its decay is required for a self-consistent theory.

The five W boson decays recorded by the UA1 detector in the 1982 run were only just consistent with this prediction. What would the improved statistics yield? This decay analysis can be extended to link the measured distribution in the angle between the electron and the proton beam to the spin of the W boson. The measurement of this distribution in W decays allows a direct measurement of the W boson spin, which was predicted to be one, the same as the photon.

In order to extract the rare energetic electron candidates from the prolific background of pions, several selections were again necessary on the sample of 115000 express line events, which had a localised transverse energy of more than 15 GeV in the electromagnetic calorimeter. An energetic charged track in the central detector pointing to the energetic calorimeter cluster was required, which reduced the sample to 18000 events. In order to minimise the risk of other particles faking the energetic electron, this track and associated calorimeter cell was further required to be relatively isolated from other energetic activity, which left just 1500 events. Finally the sample was reduced by requiring the energy deposition in the electromagnetic and hadronic calorimeters to be consistent with an electron. These 346 events were individually examined by physicists on the interactive display monitors in order to verify that they were genuine antiproton–proton collisions and that no serious reconstruction errors had occurred.

Most of these events contain at least one jet of localised calorimeter energy in addition to the electron candidate, but 50 of the events have no such activity. These two classes of events exhibit very different properties. Whilst events with jet activity show no evidence for missing transverse energy, the other events all have large missing transverse energy as shown in figure 13.1. This suggests the presence of an energetic neutrino, which is expected if these are W boson decays. What a difference to 6 months earlier when there was only a handful of such events.

Consider the production and decay of a W^- boson where it is expected that the decay electron will be emitted most frequently in the direction of the proton beam. We can understand this very striking prediction by considering the properties of the weak inter-

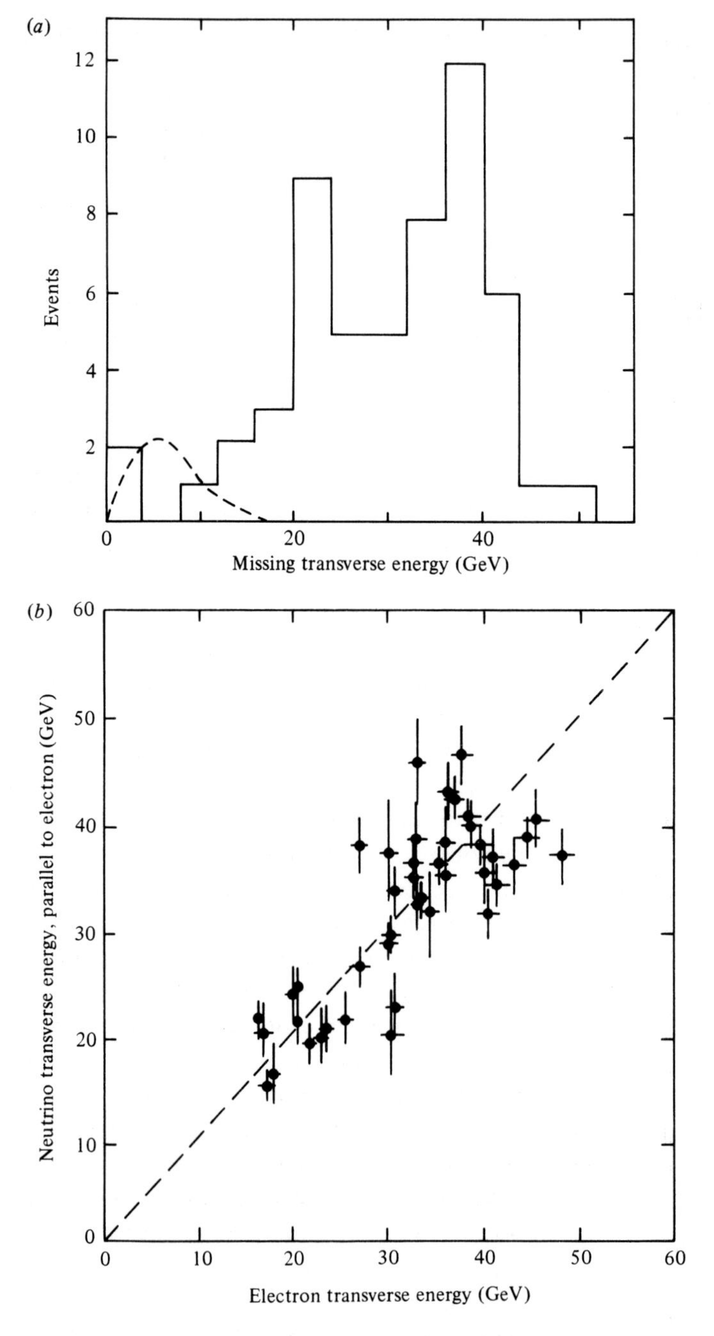

Figure 13.1. (*a*) The distribution of missing transverse energy recorded by the UA1 detector in a sample of 55 candidates for the decay of W boson into electron–antineutrino or positron–neutrino. The dotted line indicates the distribution that would be expected due to fluctuations of energy deposition in the calorimeters. The events clearly have a very significant

action and the spins of the quarks and leptons. At high energies the W boson only interacts with left-handed leptons and quarks, or right-handed antileptons and antiquarks. We recall that a left-handed particle is one for which the spin vector points in the opposite direction to its motion. The W^- boson is generally produced in a collision between a quark, from a proton, and an antiquark, from an antiproton. Because the W boson only interacts with left-handed particles at high energies, it must be produced with its spin as shown in figure 13.2*a*, because there are no other contributions to angular momentum along this axis. Now consider the two extreme W boson decays for the electron and antineutrino, shown in figure 13.2*b* and *c*. The decay shown in figure 13.2*b* is forbidden by the conservation of angular momentum but the decay with the electron in the same direction as the incident quark (or proton), as shown in figure 13.2*c*, is allowed. This asymmetry in the decay of the W boson becomes less dramatic for decays which are at an angle to the beam, but it is still very large. The direct measurement of this asymmetry measures the spin of the W boson and also demonstrates this unusual handedness in its interaction with quarks and leptons.

In order to check the angular distribution with respect to the protons, of electrons in the decays of the W boson in these events, we need to make even stricter selections. The electromagnetic energy must be deposited away from the small vertical gaps in the calorimeter and the charge of the particle needs to be reliably measured. This leaves 29 events, 20 electrons and 9 positrons, which can be used for this crucial measurement of the W boson spin. The measured angular distribution is shown in figure 13.3 and is a strongly asymmetric distribution as predicted by the electroweak theory. The curve superimposed on the data shows that the measurements strongly favour the expected spin of one for the W boson and that the W boson does indeed only interact with left-handed quarks and leptons at high energy. This very important result was presented at Brighton and was subsequently confirmed by the UA2 experiment.

We have already seen that the experiments had identified the first observed examples of a Z^0 boson decaying into an electron and a positron. The UA1 collaboration had also detected an example of

Caption for fig. 13.1 (*cont.*)
missing transverse energy which can be shown to be consistent with the presence of an energetic neutrino. (*b*) The missing transverse energy parallel to the electron is plotted against the transverse energy of the electron for the 43 W candidates which remain after the removal of events with the electron near the vertical gap in the calorimeters. The strong correlation between the two transverse energies is expected for the decay of a W boson into two particles (from G. Arnison *et al.*, *Phys. Lett.* **129B** (1983) 273).

(a)

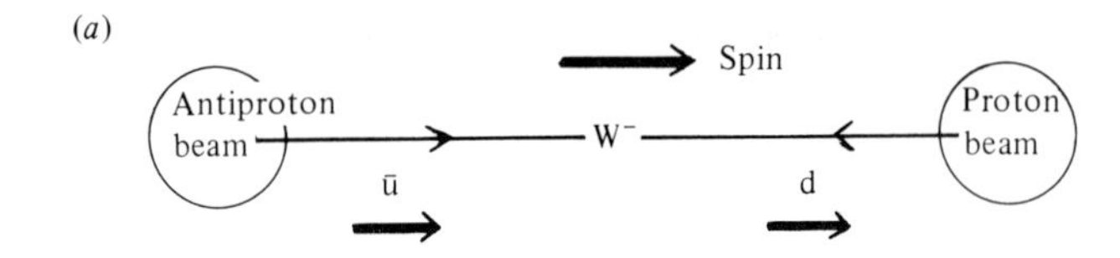

(b)

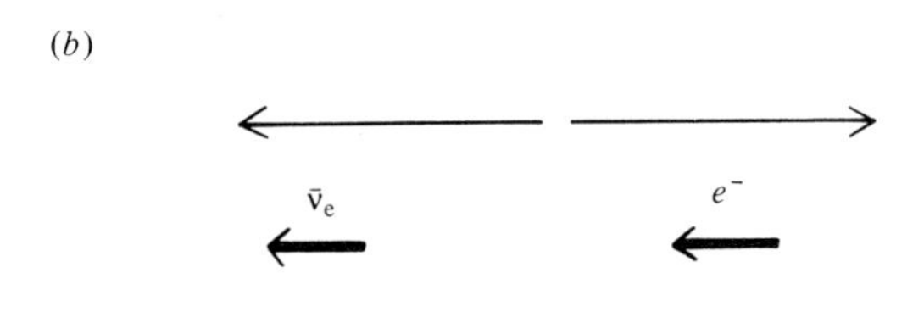

(c)

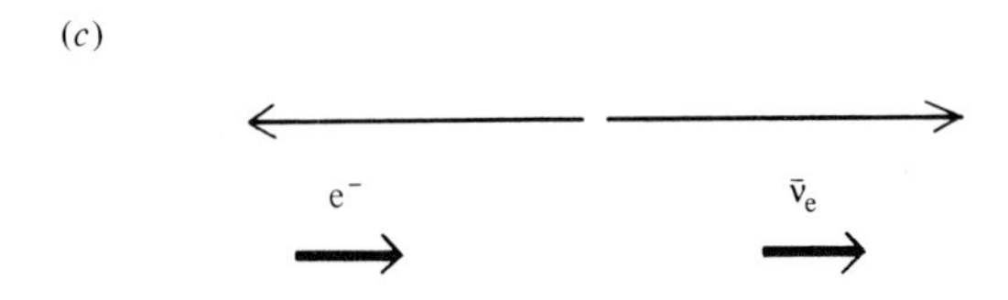

Figure 13.2. (*a*) Schematic of the production of W^- bosons in an antiproton–proton collision. At high energies the W^- boson can only be produced from right-handed antiquarks and left-handed quarks. The W^- boson will be most frequently produced from a collision between an $\bar{u}$ antiquark (from an antiproton) and a d quark (from a proton). Consequently the W^- boson spin will be in the direction of the incident antiproton beam. (*b*) This type of W^- decay is forbidden because the overall spin of the two decay products is in the wrong direction. In practice the decay particles can only be detected at a small angle to the beams, but these decays are also strongly suppressed. (*c*) This decay configuration means that the final spins combine to be in the same direction as the W^- boson and so this decay is allowed.

a Z^0 decay to muon–antimuon. The speedy identification and study of the W and Z bosons by the UA1 and UA2 collaborations provided the highlights of the Brighton conference.

There was still more to learn about the W boson. Did it also decay to a muon as expected? If so, the UA1 detector was the only one able to identify this decay. The reconstruction of the muon events was started as soon as the 1983 run finished, using a limited reconstruction method which reduced the analysis time by a factor of 50. The central detector information was processed only in a limited volume near the muon track and full processing was completed only if there was evidence for a fast charged track in this region. The W bosons are

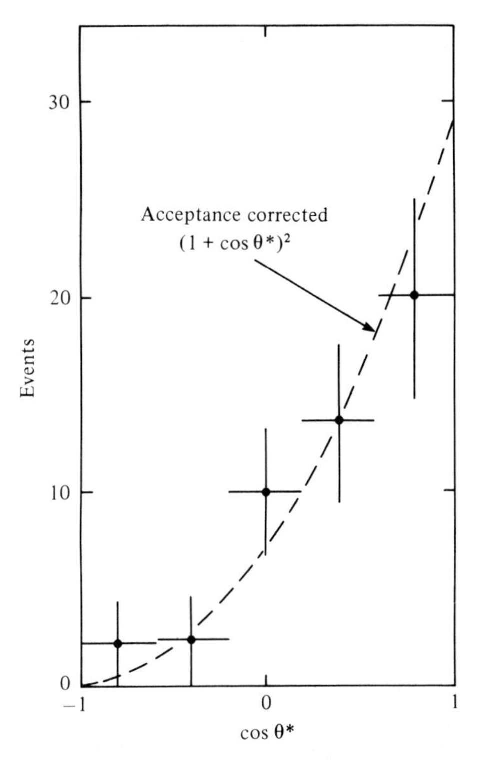

Figure 13.3. The crosses are the UA1 experimental measurements of the decay angular distribution of the electron (positron) from W^- (W^+) decay. The distribution shows the number of events (N) detected at various values of θ^*, which is the angle between the outgoing electron and incoming proton (or positron and incoming antiproton) in a system where the W boson is at rest. The dramatic asymmetry between the two sides confirms the predictions explained in figure 13.2 and the good agreement with the dotted line indicates that the W boson has a spin of one, like the photon. 'Acceptance corrected' means that the ideal theoretical curve has to be corrected to allow for the fact that a real detector can never measure particles in all directions.

expected to decay at equal rates into electrons and muons, because studies of the weak interaction have shown that there is a universal interaction strength for all leptons. However, in practice the design of the detector meant that the relative numbers of W decays into electrons and muons was difficult to measure precisely. The electronic triggers used for electron and muon events were quite different and also the muon chambers, which were used in the trigger, only covered a fraction of the area of the electromagnetic calorimeter. Finally, for a muon the quality of the central detector track has to be extremely high to exclude the possibility of another particle decaying at a small angle to a muon. For an electron, the enormous energy deposited in the calorimeter means that the track reconstruction is less critical. When all these factors were carefully evaluated, three times more electron decays were expected than muon decays of the W boson. The search for the muonic decay of the W boson yielded 14 clear candidates for the decay of the W^- boson into muon antineutrino and W^+ decay into antimuon/neutrino. This was consistent with expectations and these results were submitted for publication in December 1983.

The spin of the W boson had now been measured to be one and it only interacted with left-handed quarks and leptons. Although the statistics were small the W boson also seemed to decay at equal rates to electrons and muons.

13.3 The masses of the W and Z bosons

The electroweak theory predicted the existence of W^+, W^- and Z^0 bosons, which had now been observed, but a more detailed check on this theory is provided by the measurement of the masses of the W and Z bosons. The predicted masses depend on the gauge principle, which was the basis for the unification of the electromagnetic and weak nuclear forces, and this even predicts the exact form of the self-interaction between the W and Z bosons. These interactions, together with the interaction with the undiscovered Higgs boson, produce a shift of several GeV on the predicted masses of the W and Z bosons. Detailed measurements of these masses provide a check of these important ideas.

We have described the experimental identification of the W and Z bosons, but what about their precise mass determination? The calibrations have been described and were performed very carefully, but with a complex detector containing hundreds of separate calorimeter cells it is a formidable task to obtain absolute energy measurements to an accuracy of less than a few per cent. Ideally each cell should be inserted in a test beam at regular intervals and the response to a known energy checked at all angles and positions over the cell. This was not possible but many other calibrations were used and these have already been described.

When the analysis of W and Z bosons detected by the UA1 and UA2 experiments was completed, over 100 vector bosons had been identified. The numbers of events used in the determination of W and Z boson masses, from both experiments, are listed in figure 13.4. The experimental mass values were first determined for each decay mode, by each collaboration separately, with an uncertainty calculated from the measurements and the systematic uncertainties in the

UA1	W →	e ν	43
UA2	W →	e ν	37
UA1	W →	μ ν	14
UA1	Z →	$e^+ e^-$	4
UA2	Z →	$e^+ e^-$	4
UA1	Z →	$\mu^+ \mu^-$	5

Figure 13.4. A summary of the number of examples of W and Z boson production, detected by the UA1 and UA2 experiments in 1983, which were used in the determination of W and Z boson masses.

detector calibrations. These independent mass estimates were in very good agreement and so an overall experimental average, weighted to include the different errors, was calculated for the W and Z bosons and these are shown in the table below.

The theoretical prediction for the W and Z masses depends on one free parameter of the electroweak theory, namely $\sin^2\theta_W$, as described in section 4.7. This parameter has been measured in many independent experiments and these yield a result of $\sin^2\theta_W = 0.220 \pm 0.006$. The exact masses also depend very slightly on the undetermined masses of the top quark and Higgs boson, but by assuming reasonable values of 40 GeV and 100 GeV for these, the predicted masses for the W and Z bosons are as shown in the table below. The experimental mass for the W boson of 82.1 ± 1.7 GeV agrees very well with its predicted value of 82.4 ± 1.1 GeV. Similarly, for the Z boson, the experimental mass of 93.0 ± 1.8 GeV is in excellent agreement with the theoretical prediction of 93.3 ± 0.9 GeV. The detailed predictions of the electroweak theory, which unifies the electromagnetic and weak nuclear forces, had been verified. These measurements closed a very important chapter in the improved understanding of the fundamental forces. The W and Z bosons do exist at their predicted masses.

	W bosons			Z bosons		
	UA1 W→eν	UA1 W→μν	UA2 W→eν	UA1 Z→ee	UA1 Z→μμ	UA2 Z→ee
Number of events	43	14	37	4	5	4
Experimental mass		82.1 ± 1.7			93.0 ± 1.8	
Mass predicted from electroweak theory		82.4 ± 1.1			93.3 ± 0.9	

13.4 The search for the top quark

While the W and Z bosons were still being analysed a new analysis effort was started, in search of the top quark. This was the only missing particle in the first three generations of leptons and quarks and it had been the target of many searches for several years. The heaviest known quarks, the charm and bottom quark, have both been intensively studied at electron–positron colliders. If the electron and positron beam energies are large enough to create the quark–antiquark pair, then this is the ideal way to study the properties of the quark. As the charm and bottom quarks have masses of 1.5 and 5 GeV respectively, electron–positron colliders with energy of just 10 GeV are able to study these quarks. The mass of the top quark is not predicted in any theory, although it cannot be too heavy without leading to some theoretical problems. The electron–positron accelerators at Hamburg and Stanford are able to investigate collision energies up to 46 GeV and so can study the top quark if it has a mass of less than 23 GeV. Several years of careful study with these accelerators have shown that the top quark mass must be greater than 23 GeV.

If the top quark mass is less than the mass of the W boson, it can then be produced in W or Z decays, but we will concentrate on the more numerous W decays for this discussion. We have seen that W bosons decay equally to all available lepton and quark pairings, provided we remember that each quark decay is enhanced by a factor of three, because each quark exists with three different colour charges. The W^+ boson can in principle decay to any quark–antiquark pair with an overall electric charge of one. However, from our knowledge of the weak nuclear force, these decays are usually expected to produce a quark and an antiquark from the same generation. This means that the quark decays of the W^+ boson are expected to be dominated by $u\bar{d}$, $c\bar{s}$ and $t\bar{b}$ respectively. For every single decay of W^+ to $e^+\nu_e$, there should be three decays of W^+ to $t\bar{b}$. In addition to the 43 W bosons decaying to positron (or electron) there should be another 130 W bosons decaying to the top quark (or its antiquark)! This sounds very promising, but remember that even if these events have been recorded, they are mixed in amongst millions of events on the magnetic tapes.

How could we distinguish these events with top quarks from the rest of the data? We know that the top quark mass is greater than 23 GeV and it must be slightly less than the W mass of 80 GeV for this decay to be possible. In order to simplify the discussion, we shall assume that the unknown top quark mass is 40 GeV, although similar arguments apply whatever the mass. If the W boson is produced at rest in an antiproton–proton collision, it must convert its 80 GeV of mass energy into the motion and mass energy of its decay products. This will always result in a back to back decay configuration as in the decay to an electron and an antineutrino. In the decay of the W^+ to $t\bar{b}$, more energy is required for the quark masses, but the quark and antiquark are still given a large amount of energy for their motion.

These quarks do not appear directly as stable particles but become 'dressed' and appear in the detector as hadrons. The hadrons from the $\bar{b}$ antiquark will be collimated into a localised jet of particles because its momentum is large compared to its mass. The top quark has to carry the same momentum and as it is much heavier it moves much more slowly. When the top quark converts into hadrons, they can be scattered over a large volume of the detector and will not appear as a localised jet because of the large top quark mass. Ultimately this signal alone is not characteristic enough to isolate the top quark, as there are very large numbers of events with jets observed in the data.

Most of the events recorded on tape contain particles which are made up of up and down quarks, which are the constituents of protons and everyday matter, and in these events an electron or muon is produced only very rarely. Many of the particles produced are pions, which are also made up of up and down quarks and antiquarks. These do eventually decay to muons but usually only after the pion has moved well away from the collision point. When any of the heavier charm, bottom or top quarks are produced one in every five events contain an electron or muon and so the selection of electron or muon candidates certainly enhances the sample containing heavy quarks. These heavier quarks decay to leptons much more rapidly because of their larger mass and these leptons appear to come directly from the collision point. However, the remaining events are still dominated by the charm and bottom quarks and background events where the lepton is not genuine, so how can these be eliminated?

We can use the very different masses of the quarks to help in this separation because when the heavy quark decays to a lepton it always produces some extra hadrons, which usually collect into a jet. For this discussion we shall concentrate on events where the quarks are produced with high energy. When the charm or bottom quark decays, the electron or muon cannot escape from this accompanying jet of particles because all the decay fragments are thrown forward. The top quark is so heavy that even though it is travelling relatively

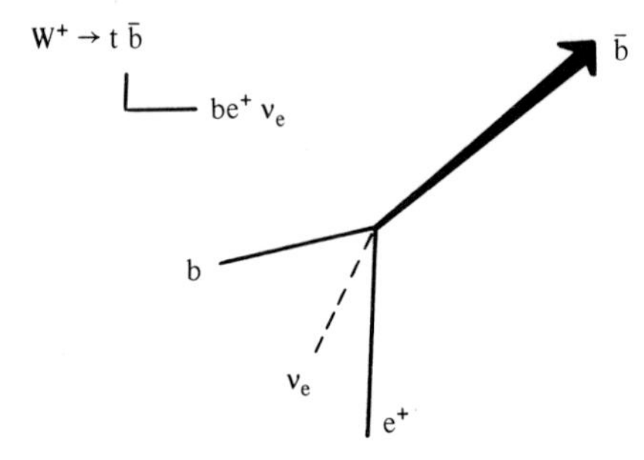

Figure 13.5. A schematic view of the decay of the W boson to a top quark (t) and a bottom antiquark ($\bar{b}$). The large mass of the W boson means that the bottom antiquark will be energetic and will balance the momentum of the top quark. In about 10% of such events the top quark is expected to decay into a bottom quark, positron and a neutrino. Owing to the large mass of the top quark these three decay fragments can be thrown into different parts of the detector and this can leave an isolated positron in addition to two observable jets.

rapidly, it can still throw its decay products in all directions and sometimes these will be well away from each other. Only the top quark can produce events with jets and isolated electrons or muons! The expected features of the decay of a W boson to the top quark are illustrated on figure 13.5.

The search for the top quark in the antiproton–proton collider made use of these simple ideas, and events with some jet activity were selected if there was evidence for an energetic electron or muon. Further selections required that the lepton in the event was very well isolated, which dramatically reduced the number of candidates, and the remaining muon–electron events were subjected to rigorous selections in order to minimise the background contamination of the sample. This search was carried out by many people in the collaboration, often trying slightly different methods to isolate the top quark. The analysis was even more difficult than the earlier extraction of the W and Z bosons and the work continued through the winter of 1983.

By the summer of 1984 the electron and muon searches were completed and after many months of careful checking, a handful of candidate events were extracted. There were three isolated electron events and three isolated muon events, each with two extra jets. One of these events is shown in figure 13.6. These events were consistent with the decay of the W boson to a top quark, but could they be produced by some other background process? When so few events are selected from such a large sample, the background calculations are very important. Could these be background events, which were

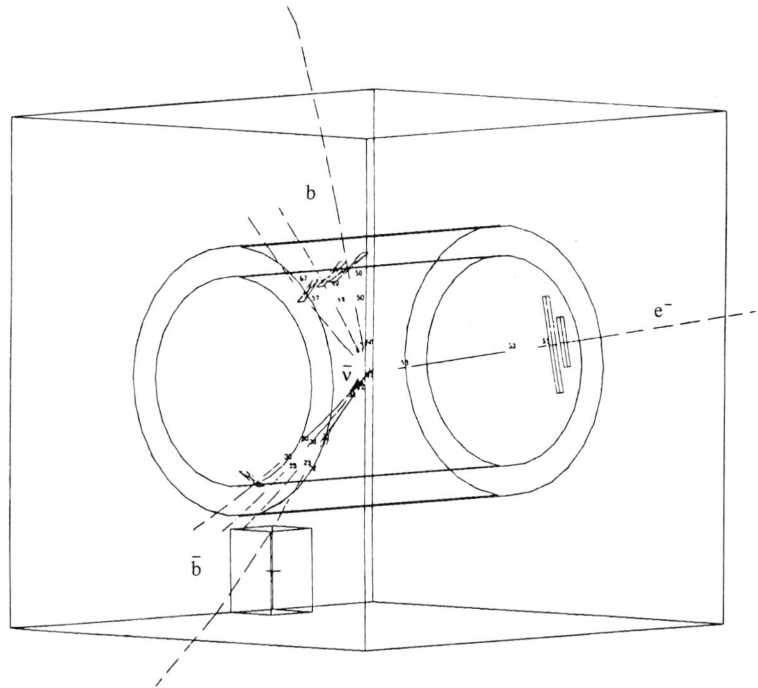

Figure 13.6. An event detected by the UA1 experiment containing an energetic isolated electron and two well-separated jets of particles. The lines correspond to the paths of charged particles and the symbols indicate where energy has been deposited in the calorimeters. This event is quite consistent with the decay of a W boson to a top quark and bottom antiquark with the subsequent decay of the top quark to an electron (from G. Arnison *et al.*, *Phys. Lett.* **147B** (1984) 493).

somehow surviving all the checks to exclude them? An intensive study of the backgrounds for electrons and muons in such events was carried out using experimental and computer-generated data. Several independent analyses were performed and these all predicted that the expected background was very small. This result was made more convincing by careful study of the expected properties of background events. Not only were there too few to explain the signal, but the event characteristics were also very different.

These isolated lepton events with two jets were inconsistent with known processes involving lighter quarks. If these events result from the decay of a W boson to a top quark, then the top quark mass must be between 30 and 50 GeV. A more precise mass determination requires more statistics and an improved measurement of the jet energies. On 3 July 1984 Michel Della-Negra presented these latest UA1 results to a packed CERN auditorium. A few days later CERN issued a press release containing the news that events consistent with the decay of the W boson to the top quark had been observed in the UA1 experiment.

13.5 More surprises?

The W, Z and top quark searches were all made by using some predicted property of the events in order to extract a small signal from an immense background. This selection has to be present in the electronic trigger and then again in the analysis of the recorded events. For a totally unexpected discovery there are clearly two serious problems. The events may not have been selected by the electronic trigger and will not then be on the magnetic tapes at all. This is unlikely, as the trigger was deliberately chosen to include as many of the energetic events as possible, independent of their detailed properties. More serious is the difficulty of extracting unusual events from the large number of recorded events as there are so many possibilities.

There were several searches for unusual events in the collider data, but we will concentrate on just one of these, to give the flavour of how these searches were conducted. We have already described the extraction of electron and muon events and these were frequently found to be accompanied by a large missing transverse energy. As most of these events appear to be decays of the W boson, this is presumably due to the emission of a neutrino. What happens if we search for events with a large missing transverse energy without any other requirements? This will be a search for neutrinos or any other particles that do not interact in the detector. The neutrino is the only neutral, non-interacting particle that has been observed in current experiments, but perhaps the very high collision energies of the collider could be producing for the first time a heavier particle with properties like the neutrino.

In the summer of 1983 a search was started for these 'missing transverse energy' events, which was even more difficult than the search for the top quark. The other selections were all based on the identification of a well-measured central detector track, calorimeter energy deposit, muon track or a jet of particles. Here the selection was based on the detection of an energy imbalance between the various calorimeter cells of the detector. Many of the events observed to have such an imbalance are due to some weakness of the detector or a background process and not to the presence of an energetic neutrino or exciting new particle. Sometimes part of the detector is hit from the outside by a particle travelling outside the beam pipe, or a cosmic ray which deposits a large amount of energy unrelated to a genuine collision. This will clearly produce a large fake missing transverse energy. The calorimeters have small gaps at the top and bottom, so energy can leak out causing an apparent imbalance. A single calorimeter cell can be struck by more than one energetic particle in the same event, and if the cell is large this can confuse the reconstruction of energy flow and lead to errors in calculating the missing energy. Finally, the statistical nature of the energy deposition in the calorimeter means that there are frequently small imbalances in transverse energy flow even in normal events. There were thousands of candidate events to study and these were methodically scanned in several of the collaborating laboratories. Gradually most of the common problems were understood and precise selections were used, in computer programs, to eliminate the majority of the background events. The most interesting candidates were those with very large missing transverse energies, where the measured effect was much larger than could be caused by any reasonable calorimeter fluctuation.

Once all the background events had been excluded, the properties of the remaining events, with a missing transverse energy exceeding 40 GeV, were studied with great interest. This sample was dominated by the W boson events that were already well known, but in addition there were a handful of new events where the missing transverse energy survived all the scanning and technical checks. Many of these events contained a single, well-collimated energetic jet of particles (monojet) which was opposite to the 'missing transverse energy'. Careful studies of the events at slightly lower missing energy were also performed, to enable the background to these dramatic events to be calculated. These calculations showed that the background for these rare events was very low from other known processes, so what were these events?

There are several possible explanations and we shall only discuss a few of them. We have already identified the separate W bosons decays to electron and muon, but what about the equal number of W bosons that are expected to decay to a tau lepton and its antineutrino? The tau lepton frequently decays to hadrons and so perhaps the single jet is the remnants of the tau lepton? However, the missing transverse energy observed in the monojet events seems to be too large for this decay so this is unlikely to be the source. The Z^0 boson can decay to a neutrino–antineutrino pair and this is expected to occur three times as often as the decays to electron–positron and muon–antimuon. Normally the Z^0 boson is produced without much sideways momentum and so when it decays into neutrinos, these will balance each other, leaving no signal in the detector or any missing transverse energy. Occasionally the Z^0 is produced recoiling sideways from a jet, as we saw in the first Z^0 to muon–antimuon decay. In this case the jet would be seen and the neutrino–antineutrino pair could produce a missing transverse energy opposite to the jet, much as observed. However, in the monojet events the jet energies are very much larger than observed in other types of Z^0 decays, but the statistics are still rather low. Are there any other possible explanations?

One of the most exciting interpretations of these spectacular events is that they are the first observed examples of the production and decay of massive supersymmetric particles. If these ideas are correct, then the missing transverse energy might be carried out of the detector by a neutral non-interacting particle called the photino. We will briefly discuss this more speculative idea in the final chapter.

13.6 The analysis of the data continues

We have concentrated on the analysis of a few of the most interesting events collected from the collider runs but there are many other studies and results that we have not included. Once the data have been successfully recorded, they can provide useful information for many years to come. The collider data have already been a gold mine of interesting results.

The theory of quark forces, quantum chromodynamics (QCD), has always been a difficult theory to compare quantitatively with experiment. Frequently such comparisons require many assumptions and this always reduces their significance. The comparison of QCD predictions with the detailed production properties of the W and Z bosons has already been very successful. Studies of the jets produced in antiproton–proton collisions have also revealed that the strong nuclear force between quarks follows an inverse square law at small distances. These are just two of the many unexpected ways in which the collider data has contributed to the general progress in our understanding of the basic forces.

Part Five

The future

Chapter Fourteen

What next?

14.1 Introduction

We have nearly completed our story but before we do let us look to the future. In the previous chapters we have briefly reviewed the important results from many particle physics experiments which have investigated the basic structure of matter and the fundamental forces since the beginning of the twentieth century. There have also been an enormous number of important theoretical ideas over the same period. This powerful combination of an experimental measurement being used as a test of a theoretical idea has led to great progress in our knowledge of the world within the atom and in all science.

The main topic of the book has been a detailed description of the experimental techniques and equipment used in a recent particle physics experiment. This international project, which was carried out at CERN, identified the W and Z bosons. These were both predicted by the electroweak theory which unifies the electromagnetic and weak nuclear forces. The success of the project was due to the efforts of many people who worked tirelessly towards its successful completion.

The special contributions of Carlo Rubbia and Simon van der Meer to this project were recognised on 14 October 1984 when they were jointly awarded the Nobel Prize for Physics. The citation from the Nobel committee ended with the words 'for your decisive contributions to the large project which led to the discovery of the field particles W and Z, communicators of the weak interaction'.

At present energies the quarks and leptons seem to be structureless, indivisible and linked together in three generations of fundamental particles. All of the known particles are either leptons or hadrons, which are composed of quarks, and the forces between them are carried by the photon, gluon, W and Z bosons. The successful unification of the electromagnetic and weak nuclear forces leaves just two other distinct fundamental forces, the strong nuclear force between quarks, which is also described by a gauge theory, and the very weak gravitational force. Current experimental and theoretical developments are continuing to probe the structure of matter and trying to improve our understanding of these fundamental forces.

The most important new discoveries will probably be complete surprises. However, we will continue this story with a brief look at a few topics in experimental and theoretical particle physics which could provide the next breakthrough in our understanding of the atom and its forces. We will first follow what has been happening at the CERN collider project since the historic 1982 and 1983 data-taking runs and see what improvements are planned for the future.

14.2 The next collider run

The next data-taking run at the CERN antiproton–proton collider was not scheduled until September 1984. The intervening year allowed several limited improvements to be made to the detectors and the UA1 collaboration concentrated on three separate areas. Extra chambers were added to increase the area covered by the muon detector system, and these included a new type of chamber that was installed inside the iron shielding. The data acquisition system was extensively improved in order to reduce the losses due to computer dead time at high collision rates. For the first time, the express line flagging was also extended to allow the selection of events with muons as well as electrons. Finally, the construction of a high-precision drift chamber to surround the beam pipe and fit inside the central detector was started. This was designed to improve the identification of particle decays occurring very close to the collision vertex. Most of the improvements to the detector were ready for the start of data taking in September 1984, but the high-precision vertex chamber had not been fully tested and was not available at this time.

What had been done to improve performance of the accelerators for collider operation? The cooling requirements of the accelerator magnets had imposed the original 270 GeV beam energy limitation, but after the experience gained in the earlier runs and the addition of extra magnet cooling this could safely be raised to 315 GeV for the 1984 run. This increase in collision energy from 540 to 630 GeV was a very useful bonus, as it increases the rate of W and Z boson production. A number of modest improvements to the antiproton accumulator and other parts of the accelerator complex were expected to increase the average collision rate by a factor of two.

The 3-month period of collider running finished on 20 December 1984. Over this period the UA1 experiment detected nearly three times as many antiproton–proton collisions as obtained in the 1983 run. A number of stored beams were lost, mainly due to power supply failures, but several beam intensity records were also broken, with one 24-hour period in 1984 yielding as many collisions as a month of the 1982 run!

The spectacular events were still appearing in these data. The W bosons were found so frequently that they no longer caused much comment, but each new Z boson was still rare enough to generate a short period of excitement. As both electron and muon events were selected for rapid analysis, the scanning area was even busier than usual. Much of the current interest centres on the study of the unusual events, which have still to be properly understood. The detailed studies of these new data are now in progress and there are plenty of interesting events to analyse.

14.3 New developments at the CERN collider

Higher-energy collisions

As previously stated, the accelerator magnet cooling currently restricts the collision energy to 630 GeV at the collider. This limit is extremely frustrating since cosmic ray interactions with unusual properties have been observed at energies a little higher than this. In order to investigate higher-energy collisions it was decided to operate the collider at energies of up to 900 GeV, but only for very short periods. The beams of antiprotons and protons were injected into the super proton synchrotron (SPS) where they are normally accelerated to 315 GeV and maintained at that energy. In the higher-energy running the beams are accelerated up to 450 GeV and kept at that energy for a fraction of a second and then decelerated again. This procedure is repeated at 10-second intervals and allows data to be recorded at several different collision energies up to 900 GeV. This novel collider operation was very successful and the UA1 and UA5 detectors recorded data under these conditions in March 1985.

This highly complex operation only yields a small number of collisions and so it is not useful for identifying W or Z bosons or other rare processes. However, it is an ideal way of making a rapid survey of the collisions at even higher energies. If the unusual effects observed in high-energy cosmic ray interactions are caused by protons, then these should certainly be observed in the two detectors at the collider. The first analysis of this data-taking run is only just beginning.

Increased interaction rate

The only way to improve significantly the collision rate of the CERN antiproton–proton collider is to upgrade the antiproton accumulator, which at present collects and cools the antiprotons. In the summer of 1984, CERN approved a project to provide a separate collector and cooling system for antiprotons (ACOL), which is planned to be completed in 1986 and should increase the collision rate by a factor of 50. This will allow high-statistics studies of W and Z bosons to be made, as well as exploratory searches for particles which are produced even less frequently.

The current detectors could not make full use of such high interaction rates and so major improvements are planned for both the UA1 and UA2 detectors when the ACOL project is completed. These detector changes are still at an early stage but the UA1 central calorimeter will be converted to use uranium instead of lead and iron because this has a more uniform response to charged and neutral particles. In the new calorimeter, cell sizes will also be smaller in order to reduce the chance of more than one particle striking the same cell. This will be essential with a higher interaction rate, and a new electronic trigger is being constructed to make use of the extra information provided by these cells.

During the 1984 run many people were involved in the design of this improved detector. It is important for some people to be planning for the next experiment, because it will take several years to construct. The manpower and effort has to be continuously balanced between analysis of older data, urgent current activities and plans for any future experiment. The optimal allocation of resources can be difficult to achieve, but in general it is handled very well in particle physics experiments. The future runs of the collider could yield even more exciting results than those of the past few years but we shall have to wait and see. Whatever happens, the early collider programme is going to be a hard act to follow!

14.4 Interesting experimental searches

In the development of science, crucial contributions have often been made by totally unexpected experimental discoveries. However, it is still interesting to speculate on the important experimental discoveries that might be made in particle physics in the next 5 years. In this brief review we will concentrate on just three areas of current experimental interest. We start with searches for new high-mass particles, including the Higgs boson. Then we discuss the experiments which are being conducted to check whether the neutrinos have a small mass and whether one neutrino type can ever change into another. Finally we introduce the underground experiments which are being conducted around the world to investigate whether the proton decays!

Higgs particle

The neutral Higgs particle is probably the most important object requiring investigation with the higher-energy electron–positron accelerators, which should be available before the end of the decade. It is very important to check whether this particle exists and if it does, what its characteristics are.

One of the basic features of all the interactions that we have discussed is that they exhibit local gauge invariance, which was introduced in chapter 4. When this idea was first applied it could not be used in cases where the force carriers were massive. Spontaneous symmetry breaking was one method introduced to solve this problem and one consequence was that a particle, named after Peter Higgs of Edinburgh University, which interacts with all the fermions and bosons, should exist. The Higgs particle provides the mass of these particles in a special way which does not break the overall symmetry. From the electroweak theory all of the Higgs properties are known except, unfortunately, its mass. As its mass could be extremely large, it could be outside the range of accelerators currently under construction. It also might not exist and the symmetry breaking may be caused by some as yet undiscovered phenomenon. This is why the experimental measurements are so important, in order to distinguish between possible theoretical explanations.

If the Higgs particle is identified and its mass is measured, then all the detailed predictions of the electroweak theory can be checked experimentally. The Higgs particle will decay to the most massive objects that are energetically allowed and this emphasises its vital role in providing mass to all particles. At the antiproton–proton collider it would be expected to be produced with a W or Z boson only once in every 1000 such events. At this low rate its detection looks very difficult, but it should be quite possible at the large electron–positron accelerator, as mentioned in chapter 7, where 10000 Z bosons per day are expected!

Properties of neutrinos

One of the most interesting particles that have ever been discovered is the mysterious neutrino, which interacts very weakly and seems to have a mass of zero. It is one of the decay products of the W boson and so it has already appeared several times in our story, but we have not been able to study its properties in any detail. We have already met the three types of neutrino, ν_e, ν_μ and ν_τ which seem to be different particles. If all types of neutrino have a mass of exactly zero, then the three different types can always remain distinct. However, if the neutrinos have a small mass, which differs for each type, then the phenomenon of neutrino oscillations becomes possible and an electron-type neutrino could change into a muon-type neutrino and vice versa. If the neutrino has a small mass this could also have major implications for both particle physics and cosmology. Several experiments are in progress throughout the world, searching for any evidence of neutrino oscillations. These experiments are being conducted at nuclear reactors and accelerators but none of these has produced any conclusive evidence for this effect. There are also experiments being designed to extend this search to the neutrinos that arrive at the earth from the sun. As it takes 10 minutes for them to make this journey, these experiments would allow a search for neutrino stability over a much longer distance than would ever be possible on earth. The neutrinos are the only known particles which can carry information from the interior of the sun or stars without interacting and so these neutrino experiments could provide some very interesting new experimental data.

The mass of the electron-type neutrino can also be measured directly in a laboratory from very precise studies of radioactive β decay. A very detailed experiment of this type has been carried out over the past few years by a team of physicists at the ITEP laboratory in Moscow. Measurement of a very small mass requires very careful experimentation, clever design and a detailed knowledge of the atomic physics of the problem. The experiment consists of measuring very accurately the spectrum of energies of the electrons emitted in the β decay of tritium, which is a radioactive isotope of hydrogen. In this experiment the tritium is contained in a complex atom and so the theoretical interpretation of the experiment is very difficult. However, this experiment indicates that the electron-type neutrino has a mass of more than 20 eV. If this result were to be confirmed, this would have very important consequences.

Proton decay?

Some particle physics experiments are located far from accelerators and these include the experiments which are searching for evidence of proton decay. Every atom contains at least one proton and so if the proton were to decay, this would probably be the most exciting discovery of all! Our own existence implies that the mean lifetime of the proton must exceed 10^{20} years, but we know that a mean lifetime is just a characteristic lifetime and that there is a high probability of decay in a shorter time. When the problem is studied carefully it is found that proton decay could probably be observed, even if the proton's lifetime were as long as 10^{32} years! By a coincidence some theories which unify the strong nuclear force with the electroweak interaction predict that the proton is unstable with a lifetime close to this observable limit.

There are now many experiments searching for evidence of proton decay which are being conducted at sites far underground, often in deep mines. As the experiments are trying to identify a very rare process, other more common types of interaction need to be rejected. At these underground sites most cosmic rays are absorbed by the rock and so only neutrinos are able to reach the detectors. The aim of these detectors is to identify the decay products of a proton by containing all of its decay energy and currently these detectors are of two main types. In one type of detector, a huge tank of water acts as both the source of protons and the detector. Light travels more slowly in water than in air and so it is possible for energetic particles to exceed the speed of light when they travel in water. When charged particles travel at a speed which exceeds that of light, so called Cerenkov light is emitted. This light is emitted in a cone around the particle at an angle which depends on its speed. Cerenkov light is mainly blue and this phenomenon is frequently used in experiments for particle identification. If a proton decay occurs then its decay products will be emitted at high velocity and generate Cerenkov light, which is emitted around the direction of motion of each of the particles. The decay can then be reconstructed from the information recorded by photomultipliers which are placed on each side of the tank. This detector is particularly sensitive to the decay $p \rightarrow e^+\pi^0$, which was predicted to be the dominant decay mode of the proton in some unified models. In this mode the π^0 converts to two photons, which each cause electron–positron showers in the detector, and so Cerenkov light is emitted by all the decay products.

The other type of detector incorporates a fine-grained calorimeter, similar to those used in the collider experiments, consisting of alternating slabs of absorber and detector. The detector can be a piece of scintillator, drift chamber or a streamer tube, and as the total energy of the decay products is measured it can be matched to the known mass of the proton. This method has the clear advantage that it is able to detect many different types of decay, but the protons are parts of a complex nucleus, which means that the decaying proton may not be stationary.

The larger proton decay detectors weigh many hundred tonnes and contain over 10^{30} protons and neutrons. Even if the proton lifetime is as long as 10^{30} years, these detectors should still record several decays each year, a very difficult interaction rate to the antiproton–proton collider and other accelerator-based experiments! Some of the major proton decay experiments have now been recording experimental data for several years. Many events are recorded and the difficulty is to reject all the candidate events which could be due to other background processes. The most difficult background events are those due to neutrinos which penetrate down to the underground detectors. However, these types of interaction are already well studied in accelerator experiments and so in principle these backgrounds can be removed. A few interesting candidates for proton decay have been observed in some of these detectors, but there is still no conclusive evidence that the proton does decay. The current limits on the proton lifetime are already beginning to look inconsistent with the predictions of the simplest model which unifies the electroweak and strong nuclear forces. The next few years will provide very important results on the stability of the proton, a fact which has to be explained by any successful theory of the fundamental forces.

14.5 Current theories in particle physics

The frontier developments in theoretical particle physics sometimes involve very abstract mathematics and many more ideas than can be included here. Some theoretical developments have wide application and clear experimental consequences and as an illustration we will discuss two of these ideas in this section. However, there is no guarantee that these will remain as important elements in future theories of particle structure and fundamental forces.

We will start this section with a brief discussion of grand unified theories which attempt to unify the electroweak interaction with the strong nuclear force, and go on to describe the attempts to include gravitational forces in this unification. This introduces the idea of supersymmetry, which links the quarks and leptons with the fundamental force-carrying particles – the graviton, photon, gluon, W and Z bosons. In each case only a few basic concepts are presented together with some experimental predictions. We have already seen that the electroweak theory, which unifies the seemingly different electromagnetic and weak nuclear forces, has been very successful. The new force carriers of this unified interaction, the W^-, W^+ and Z^0 bosons, have now all been identified by experiments and this success encourages us to look for even more unification between the fundamental forces. The obvious next step would be to link the strong force between the quarks, which is described by quantum chromodynamics (QCD), with the electroweak interaction. Both of these forces are described by gauge theories which were introduced in chapter 4. Briefly this means that we expect any physical measurements to be independent of the choice of phase for the wave functions of interacting particles. If we allow an arbitrary choice at all points of space and time in the universe this is called local gauge invariance. This requirement places severe restrictions on the way that the carrier of the force can interact and in fact fixes the properties of the force carriers, the photon and gluon. Although these forces are both described by gauge theories many of their properties are very different. In QCD the interquark force is mediated by a gluon, which carries colour charge, and this force depends dramatically on the quark separation. At small distances the quarks seem free, whereas at large distances the force is so strong that the quarks seem permanently confined in hadrons.

The simplest grand unified model is called SU(5), where the number equals the size of the basic group of particles in the theory. In this group, three quarks and two leptons are mixed together, and so it is natural for a quark to change into a lepton and vice versa. The unification of weak and electromagnetic forces led to the prediction of W and Z bosons with masses near 100 GeV. If this unification is extended in a similar way to the colour force between quarks, then this new grand unified theory predicts even more massive X bosons, with masses of 10^{15} GeV! These X bosons carry fractional charges of $\pm 1/3$ and $\pm 4/3$ and are so massive that they could probably only be created at the very high temperatures of the origin of the universe. At very high temperatures, or equivalently at very high energies, the electromagnetic, weak and strong nuclear forces would all have the same strength, and quarks and leptons would freely convert into one another. At the much lower energies explored by our current accelerators, the X bosons cannot be produced and so the quarks and leptons remain distinct. At still lower energies, as experienced by us, there is not enough energy to produce even the W and Z bosons and so the weak nuclear force is much weaker than the electromagnetic interaction. This illustrates the beauty of the unified theories of fundamental forces. At very high energies all of these interactions are identical but at the lower energies of our world they are very different.

Once a theory links quarks and leptons it is very likely that a proton, which is made of quarks, will decay into leptons. The grand unified theories have had many successes and explain naturally the remarkable link between the charges of the quarks and the leptons. They also predict quite accurately the value of the single electroweak parameter $\sin^2\theta_W$ which in the electroweak theory has to be measured experimentally. However, they also predict that the proton should decay at a rate which is very close to being excluded by current proton decay experiments. It looks as if this simplest version of a grand unified theory is not consistent with the facts but it is still to early to be sure.

Supersymmetric theories (SUSY) link the basic particles quarks and leptons with the force-carrying photons, gluons, W and Z bosons. When considering the unification of the electroweak interaction with the strong nuclear force we saw that force carriers called X bosons were required with masses of 10^{15} GeV. As this mass is so much larger than that of the W and Z bosons (10^2 GeV) it is very difficult to explain the enormous differences in mass of these force carriers. In supersymmetric theories this difference is explained naturally, but only at the expense of introducing new undiscovered particles. In the calculation of particle masses, contributions from fermions and bosons enter with opposite signs, and so if each known particle was accompanied by a supersymmetric partner of the opposite type of spin, a nearly exact cancellation can be achieved. In this way the remarkable difference in masses between the force carriers could be explained.

Where are these new supersymmetric particles and why have they avoided detection? The two simplest explanations are that they do not exist or have masses too large for detection at current accelerators. The masses are not predicted very accurately, but these supersymmetric particles cannot be too different in mass to the known particles, otherwise the required cancellations would not be successful. If they exist they should be accessible to current accelerators or those under construction.

If these ideas are correct, each known particle has an undiscovered supersymmetric partner. This idea may be far fetched, but do not forget the antiparticles which have played a crucial part in our story. The prediction of an antiparticle for every particle turned out to be correct. There is no single supersymmetric theory and although these ideas do not have the predictive power of the electroweak theory, they are nevertheless very interesting. Other neutral weakly interacting particles, which would escape detection, are predicted in addition to the neutrino.

A final unification of the gravitational force with a grand unified theory of electromagnetic, weak and strong nuclear forces has long been an aim of the theories of fundamental forces. There are still many problems to be overcome but the recent developments of supergravity have introduced some very interesting new ideas. Any theory of supergravity has also to be able to reproduce the outstandingly successful general theory of relativity. The electromagnetic and nuclear forces can be described by theories which allow exact calculations to be made, even in the subatomic world. However, there are no equivalent theories which allow similar calculations for the gravitational force. The very weak gravitational force, which is always attractive, is believed to be carried by a graviton which has a spin of 2. The weakness of this interaction has made searches for gravitons, or gravity waves, very difficult and no conclusive experimental evidence for the graviton has yet been detected.

14.6 Cosmology and particle physics

It seems remarkable that the studies of the very smallest and largest objects in the universe have any common features at all. However, the early conditions in the universe involved such enormously large temperatures and energies that study of the universe provides a way of testing particle physics theories at energies larger than will ever be attainable with accelerators.

The standard big bang model has been very successful in explaining the expansion of the universe and the amounts of hydrogen, helium and other elements present in the universe. In this model the whole of the present universe was created by an enormous explosion. The microwave background, discovered by Penzias and Wilson in 1965, which consists of electromagnetic radiation, with a characteristic temperature of three degrees absolute, is detected in all directions and is the present-day remnant of the original big bang. It is presently observed at this low temperature because of cooling due to the expansion of the universe. In every direction stars are moving away from us with speeds which are proportional to their distance from the earth, but we are not in a favoured location because the same observation would be made at any place in the universe. We can measure the speeds of the stars relative to us by the observed change in wavelength of known spectral lines emitted by familiar atoms. This wavelength is changed towards the red end of the spectrum for objects moving away from us.

The universe comprises stars, galaxies and clusters of galaxies but over a very large scale it is uniformly populated in all directions. This regularity is rather difficult to explain in the conventional big bang model, because many parts of the universe could never have been in causal contact. This means that even signals travelling at the speed of light could never have been exchanged between these regions, and so it is hard to see how equilibrium and uniformity could have been achieved. The most exciting breakthrough in these theories has come in the last few years with the so-called inflationary models of the universe. According to these models, there was a very rapid expansion of the universe in the first 10^{-35} seconds of its existence, which allows very distant parts of the universe to have been in causal contact during this period. After this short time a major phase transition occurred which reheated the universe to a much higher temperature and following this the expansion reverted to that of the standard big bang model. In the inflationary model the phase change involves the Higgs field, and this is the link to particle physics. Better understanding of the Higgs particle will have direct application in such models of the early universe.

Another area where the interests of particle physics and cosmology overlap is in the determination of the number of types of neutrinos. The cosmological estimates make use of the measured helium abundance in the universe. Most of the helium is believed to have been created by nucleosynthesis a few seconds after the big bang. The helium created at this time depends on the relative numbers of neutrons and protons, which in turn depends on the number of neutrino types because this affects the expansion rate of the universe. Current measurements of helium, when taken with various assumptions, indicate that there can be no more than four types of neutrino.

Measurement of the numbers of neutrinos can be made much more directly in particle physics experiments, but at present this result is less restrictive than that from cosmology. This should change with accurate measurements of the Z^0 boson which should be made within the next 5 years. As discussed earlier, each unstable particle has a spread in mass, arising from the uncertainty principle. The Z^0 boson is expected to decay into all the neutrino and antineutrino types, providing these are massless or have small masses. Each of these decay modes can be calculated to increase the spread in mass of the Z^0 boson by 0.18 GeV and an accurate measurement of this quantity should enable the precise number of light neutrinos to be determined. This will be measured in the high-energy electron–positron accelerators in the next few years.

As far as we know, the universe is composed of matter and not antimatter, as the latter has only been identified as the product of energetic collisions. It is possible that there are regions of antimatter in remote parts of the universe, but we have no evidence for any. Can we explain why the universe seems to be dominated by matter? The big bang would have been expected to produce matter and antimatter in equal quantities, so where has all the antimatter gone? In chapter 4 we saw that reactions and their 'mirror' reactions proceeded at the same rate except for those involving the weak nuclear force. A reaction is converted to its mirror reaction by the parity (P) operation and so parity is conserved by all interactions except the weak nuclear force. We can extend this idea by also changing each particle to an antiparticle (CP operation) and we find that CP is conserved by all interactions. However, even CP is violated occasionally in the decay of the neutral kaon. In some grand unified theories, this CP violation can be explained naturally and implies that the massive X bosons which mediate proton decay can cause very slight differences in the decay rates of baryons and antibaryons. In this picture, equal amounts of matter and antimatter were created by the big bang, but the slight imbalance in the decays of matter and antimatter have now led to the observed dominance of matter. There is some quantitative support for this picture because the CP-violating effects occur at a level of one in 10^8. When the number of baryons is compared to the number of photons in the universe, this ratio is also estimated to be one in 10^8 and so all the matter in the universe could be the remnant from a period when equal amounts of matter and antimatter existed. The bulk of matter and antimatter have since annihilated, leaving only this slight excess of baryons because of CP violation!

Finally, another dramatic cosmological observation – when the motion of galaxies is carefully analysed an interesting and very significant anomaly has been observed. This anomaly is present in all types of galaxy and in all parts of the visible universe. The mass of a galaxy can be deduced from the amount of visible light that is emitted and also from its rotational motion. These analyses show that there is about 10 times more mass in the galaxies than can be explained from the visible matter. This is known as the 'missing mass' problem and unseen dark matter could be present in many forms, including black holes, small planets, dust or even particles such as the neutrino, if they have a mass. There are some constraints on the form of this matter from theories of galaxy formation, but measurements of the mass of neutrinos are urgently required to help with the solution of this mystery.

14.7 Conclusion

We have now reached the end of a long story. Many of the current ideas and recent results from particle physics have been introduced in this self-contained account. It has been necessary to cover briefly the introductory concepts of particles, forces, detectors and accelerators before the main story of the discovery of W and Z bosons could be presented.

It has been very exciting to have been a part of this large and successful project and I hope that I have communicated some of this excitement to you. I have tried to describe what it was like to be a member of an experimental team at the collider. I have also tried to describe the activities of the collaboration, from the design of the detector to the extraction of the most recent results. Over the past few years I have spent much of my time in the scanning area around the display monitors, and so my snapshot of the experiment is made from this perspective. Fortunately for me, this was an area where many of the most exciting discoveries were made, and so I have been personally involved in most of the key moments of the story.

Finally I have tried to put this story into context by discussing some of the other important experimental and theoretical developments in our field. The powerful new accelerators which will be available in the next few years should complete a very successful decade of research into fundamental particles and forces.

Appendix

The title pages of the scientific papers announcing the discovery of the W boson by the UA1 and UA2 experiments at CERN

Volume 122B, number 1 PHYSICS LETTERS 24 February 1983

EXPERIMENTAL OBSERVATION OF ISOLATED LARGE TRANSVERSE ENERGY ELECTRONS WITH ASSOCIATED MISSING ENERGY AT $\sqrt{s}$ = 540 GeV

UA1 Collaboration, CERN, Geneva, Switzerland

G. ARNISON [j], A. ASTBURY [j], B. AUBERT [b], C. BACCI [i], G. BAUER [1], A. BÉZAGUET [d], R. BÖCK [d], T.J.V. BOWCOCK [f], M. CALVETTI [d], T. CARROLL [d], P. CATZ [b], P. CENNINI [d], S. CENTRO [d], F. CERADINI [d], S. CITTOLIN [d], D. CLINE [1], C. COCHET [k], J. COLAS [b], M. CORDEN [c], D. DALLMAN [d], M. DeBEER [k], M. DELLA NEGRA [b], M. DEMOULIN [d], D. DENEGRI [k], A. Di CIACCIO [i], D. DiBITONTO [d], L. DOBRZYNSKI [g], J.D. DOWELL [c], M. EDWARDS [c], K. EGGERT [a], E. EISENHANDLER [f], N. ELLIS [d], P. ERHARD [a], H. FAISSNER [a], G. FONTAINE [g], R. FREY [h], R. FRÜHWIRTH [l], J. GARVEY [c], S. GEER [g], C. GHESQUIÈRE [g], P. GHEZ [b], K.L. GIBONI [a], W.R. GIBSON [f], Y. GIRAUD-HÉRAUD [g], A. GIVERNAUD [k], A. GONIDEC [b], G. GRAYER [j], P. GUTIERREZ [h], T. HANSL-KOZANECKA [a], W.J. HAYNES [j], L.O. HERTZBERGER [2], C. HODGES [h], D. HOFFMANN [a], H. HOFFMANN [d], D.J. HOLTHUIZEN [2], R.J. HOMER [c], A. HONMA [f], W. JANK [d], G. JORAT [d], P.I.P. KALMUS [f], V. KARIMÄKI [c], R. KEELER [f], I. KENYON [c], A. KERNAN [h], R. KINNUNEN [e], H. KOWALSKI [d], W. KOZANECKI [h], D. KRYN [d], F. LACAVA [d], J.-P. LAUGIER [k], J.-P. LEES [b], H. LEHMANN [a], K. LEUCHS [a], A. LÉVÊQUE [k], D. LINGLIN [b], E. LOCCI [k], M. LORET [k], J.-J. MALOSSE [k], T. MARKIEWICZ [d], G. MAURIN [d], T. McMAHON [c], J.-P. MENDIBURU [g], M.-N. MINARD [b], M. MORICCA [i], H. MUIRHEAD [d], F. MULLER [d], A.K. NANDI [j], L. NAUMANN [d], A. NORTON [d], A. ORKIN-LECOURTOIS [g], L. PAOLUZI [i], G. PETRUCCI [d], G. PIANO MORTARI [i], M. PIMIÄ [e], A. PLACCI [d], E. RADERMACHER [a], J. RANSDELL [h], H. REITHLER [a], J.-P. REVOL [d], J. RICH [k], M. RIJSSENBEEK [d], C. ROBERTS [j], J. ROHLF [d], P. ROSSI [d], C. RUBBIA [d], B. SADOULET [d], G. SAJOT [g], G. SALVI [f], G. SALVINI [i], J. SASS [k], J. SAUDRAIX [k], A. SAVOY-NAVARRO [k], D. SCHINZEL [f], W. SCOTT [j], T.P. SHAH [j], M. SPIRO [k], J. STRAUSS [l], K. SUMOROK [c], F. SZONCSO [l], D. SMITH [h], C. TAO [d], G. THOMPSON [f], J. TIMMER [d], E. TSCHESLOG [a], J. TUOMINIEMI [e], S. Van der MEER [d], J.-P. VIALLE [d], J. VRANA [g], V. VUILLEMIN [d], H.D. WAHL [l], P. WATKINS [c], J. WILSON [c], Y.G. XIE [d], M. YVERT [b] and E. ZURFLUH [d]

Aachen [a] – Annecy (LAPP) [b] – Birmingham [c] – CERN [d] – Helsinki [e] – Queen Mary College, London [f] – Paris (Coll. de France) [g] – Riverside [h] – Rome [i] – Rutherford Appleton Lab. [j] – Saclay (CEN) [k] – Vienna [l] Collaboration

Received 23 January 1983

We report the results of two searches made on data recorded at the CERN SPS Proton–Antiproton Collider: one for isolated large-E_T electrons, the other for large-E_T neutrinos using the technique of missing transverse energy. Both searches converge to the same events, which have the signature of a two-body decay of a particle of mass ~80 GeV/c^2. The topology as well as the number of events fits well the hypothesis that they are produced by the process $\bar{p} + p \rightarrow W^{\pm} + X$, with $W^{\pm} \rightarrow e^{\pm} + \nu$; where $W^{\pm}$ is the Intermediate Vector Boson postulated by the unified theory of weak and electromagnetic interactions.

[1] University of Wisconsin, Madison, WI, USA.
[2] NIKHEF, Amsterdam, The Netherlands.

Volume 122B, number 5,6 PHYSICS LETTERS 17 March 1983

OBSERVATION OF SINGLE ISOLATED ELECTRONS OF HIGH TRANSVERSE MOMENTUM IN EVENTS WITH MISSING TRANSVERSE ENERGY AT THE CERN $\bar{p}p$ COLLIDER

The UA2 Collaboration

M. BANNER[f], R. BATTISTON[1,2], Ph. BLOCH[f], F. BONAUDI[b], K. BORER[a], M. BORGHINI[b], J.-C. CHOLLET[d], A.G. CLARK[b], C. CONTA[e], P. DARRIULAT[b], L. Di LELLA[b], J. DINES-HANSEN[c], P.-A. DORSAZ[b], L. FAYARD[d], M. FRATERNALI[e], D. FROIDEVAUX[b], J.-M. GAILLARD[d], O. GILDEMEISTER[b], V.G. GOGGI[e], H. GROTE[b], B. HAHN[a], H. HÄNNI[a], J.R. HANSEN[b], P. HANSEN[c], T. HIMEL[b], V. HUNGERBÜHLER[b], P. JENNI[b], O. KOFOED-HANSEN[c], E. LANÇON[f], M. LIVAN[b,e], S. LOUCATOS[f], B. MADSEN[c], P. MANI[a], B. MANSOULIÉ[f], G.C. MANTOVANI[1], L. MAPELLI[b], B. MERKEL[d], M. MERMIKIDES[b], R. MØLLERUD[c], B. NILSSON[c], C. ONIONS[b], G. PARROUR[b,d], F. PASTORE[b,e], H. PLOTHOW-BESCH[b,d], M. POLVEREL[f], J.-P. REPELLIN[d], A. ROTHENBERG[b], A. ROUSSARIE[f], G. SAUVAGE[d], J. SCHACHER[a], J.L. SIEGRIST[b], H.M. STEINER[b,3], G. STIMPFL[b], F. STOCKER[a], J. TEIGER[f], V. VERCESI[e], A. WEIDBERG[b], H. ZACCONE[f] and W. ZELLER[a]

[a] *Laboratorium für Hochenergie physik, Universität Bern, Sidlerstrasse 5, Bern, Switzerland*
[b] *CERN, 1211 Geneva 23, Switzerland*
[c] *Niels Bohr Institute, Blegdamsvej 17, Copenhagen, Denmark*
[d] *Laboratoire de l'Accélérateur Linéaire, Université de Paris-Sud, Orsay, France*
[e] *Dipartimento di Fisica Nucleare e Teorica, Università di Pavia and INFN, Sezione di Pavia, Via Bassi 6, Pavia, Italy*
[f] *Centre d'Etudes nucléaires de Saclay, France*

Received 15 February 1983

We report the results of a search for single isolated electrons of high transverse momentum at the CERN $\bar{p}p$ collider. Above 15 GeV/c, four events are found having large missing transverse energy along a direction opposite in azimuth to that of the high-p_T electron. Both the configuration of the events and their number are consistent with the expectations from the process $\bar{p} + p \to W^{\pm}$ + anything, with $W \to e + \nu$, where $W^{\pm}$ is the charged Intermediate Vector Boson postulated by the unified electroweak theory.

Bibliography

There are a number of books and articles which explain the important ideas in particle physics in a non technical way. In this section I have included a representative selection of these for further reading.

Background to particle physics

The Cosmic Onion: Quarks and the Nature of the Universe, F. Close. Heinemann, 1983

The Forces of Nature, P. C. W. Davies. Cambridge University Press, 1979

Superforce: The Search for a Grand Unified Theory of Nature, P. C. W. Davies. S & S, 1984

Quarks: The Stuff of Matter, H. Fritzch. Penguin, 1983

Nature of Matter, J. H. Mulvey (ed.). Clarendon Press, 1981

The Particle Play, J. Polkinghorne. Freeman, 1979

Building the Universe, C. Sutton (ed.). Blackwell, 1984

The Particle Connection, C. Sutton. Hutchinson, 1984

CERN and accelerators

Europe's Giant Accelerator, M. Goldsmith & E. Shaw. Taylor & Francis, 1977

Relativity

Relativity Visualised, L. Epstein. Insight Press, 1985

Quantum mechanics

Taking the Quantum Leap, F. A. Wolf. Harper & Row, 1981

Particle physics and cosmology

The Key to the Universe, N. Calder. Viking, 1977

The First Three Minutes, S. Weinberg. Flamingo, 1983

More advanced particle physics

The Ideas of Particle Physics, J. Dodd. Cambridge University Press, 1984
Elementary Particles, I. S. Hughes. Cambridge University Press, 1985
An Introduction to High Energy Physics, D. H. Perkins. Addison Wesley, 1982

Scientific American articles

Gauge Theories of Force between Elementary Particles, G. t'Hooft, June 1980
Cosmic Asymmetry between Matter and Antimatter, F. Wilczek, Dec. 1980
A Unified Theory of Elementary Particles and Forces, H. Georgi, April 1981
Decay of the Proton, S. Weinberg, June 1981
Search for Intermediate Vector Bosons, D. Cline, C. Rubbia, S. Van der Meer, March 1982
Structure of Quarks and Leptons, H. Harari, April 1983
The Hidden Dimensions of Space Time, D. Freedman & P. van Nieuwenhuizen, March 1985
Elementary Particles and Forces, C. Quigg, April 1985

Name Index

Subject Index